AF352926

# Freshwater Macroinvertebrates: Main Gaps and Future Trends

# Freshwater Macroinvertebrates: Main Gaps and Future Trends

Editors

**Angela Boggero**
**Laura Garzoli**

MDPI • Basel • Beijing • Wuhan • Barcelona • Belgrade • Manchester • Tokyo • Cluj • Tianjin

*Editors*
Angela Boggero
National Research Council
Water Research Institute
Verbania
Italy

Laura Garzoli
National Research Council
Water Research Institute
Verbania
Italy

*Editorial Office*
MDPI
St. Alban-Anlage 66
4052 Basel, Switzerland

This is a reprint of articles from the Special Issue published online in the open access journal *Water* (ISSN 2073-4441) (available at: www.mdpi.com/journal/water/special_issues/Freshwater_Macroinvertebrates).

For citation purposes, cite each article independently as indicated on the article page online and as indicated below:

LastName, A.A.; LastName, B.B.; LastName, C.C. Article Title. *Journal Name* **Year**, *Volume Number*, Page Range.

**ISBN 978-3-0365-2623-2 (Hbk)**
**ISBN 978-3-0365-2622-5 (PDF)**

# Contents

# About the Editors

**Angela Boggero**

Boggero Angela is researcher at the National Research Council–Water Research Institute (CNR-IRSA), Verbania Pallanza, Italy. Her work is mainly focused on the effects of human-induced impacts (e.g., climate change, acidification, eutrophication, hydro-morphology, persistent organic pollutants, and alien invasive species) on the structure of macroinvertebrate assemblages and, more in general, on freshwater native biodiversity through a macro-ecological approach. Her interests include taxonomy, biogeography, and ecology of freshwater chironomids and oligochaetes, water management, ecological bioindication and biomonitoring also through biomolecular techniques, and invasive crayfish species.

**Laura Garzoli**

Garzoli Laura is researcher at the National Research Council–Water Research Institute (CNR-IRSA), Verbania Pallanza, Italy. A biologist, she obtained her PhD in Experimental Ecology and Geobotany at the University of Pavia. She has a decade of experience as a fungal taxonomist, focusing her activity on marine and plant pathogenic fungi (mainly Pyricularia oryzae), and gaining experience in biothecnological applications and diagnostic for industries. She also works on invasive/non-native species, with a focus on alien crayfish, trying to define managing strategies for their control, and raising awareness on the issue. She is a member of the Executive Committee of INVASIVESNET–the global network of networks on Invasive Alien Species. She is curating the global survey on invasive alien species organizations and networks, contributing also to the upcoming IPBES invasive alien species assessment.

# Preface to "Freshwater Macroinvertebrates: Main Gaps and Future Trends"

Being also known as "the blue planet", Earth is almost entirely covered with water. Nonetheless, only a small percentage (2.2% of the total) is freshwater. Of this, less than a third is available for living organisms. Facing raising threats posed by global changes, freshwaters are nowadays an increasingly rare, precious, and non-renewable resource, which should not be wasted.

Macroinvertebrates are ubiquitous organisms found in both fresh and brackish waters all around the globe, from streams of different sizes to lakes, wetlands, ponds, river estuaries, and lagoons. They are often unevenly distributed and relatively difficult to sample, especially in deep sediments where they play a crucial role in linking sediments and their processes to the food webs. Therefore, their species richness and their function are mostly neglected until the ecosystems show visible environmental modifications. Indeed, several factors impact species assemblages locally and globally: nitrogen deposition, salinity and temperature increase, pollution (e.g., pesticides and heavy metals), introduction of alien species, floods or droughts.

Hence, we developed the idea of a Special Issue that dealt with freshwaters and macroinvertebrates, with the aim of describing recent developments in biomonitoring of lakes and rivers, and of identifying future scientific developments to provide data and updates to politicians, water managers, technicians of environmental agencies, and researchers. The involvement of everyone is a key factor in planning and developing future activities for its management and fair and sustainable use.

Freshwater ecology used several approaches to study and sample macroinvertebrates, to develop metrics and indices to be used to evaluate the ecological status of the environments they inhabit, to standardize the sampling and the classification of water status. Moreover, problem-solving solutions arise from the development of experimental designs useful to monitor the presence or the absence of macroinvertebrates or their abundances. This Special Issue presents the past and current knowledge on freshwater macroinvertebrates to understand their role as providers of ecosystem services, to highlight the effects of global changes on their community structure, and to underline major gaps in their study. A special emphasis is dedicated to their value as biological indicators of environmental change for the assessment of water ecological status and human risk.

The papers submitted highlight that: i) the Water Framework Directive could have a worldwide application in its general term to obtain robust and shareable data; ii) diatoms could be used in biomonitoring programs supporting researchers with information that provides a different and integrated perspective with respect to the sole use of macroinvertebrates; iii) data sharing is a useful tool to derive larger scale analyses and distribution patterns of macroinvertebrate assemblages under climatic change stress, reducing at once costs and time for river ecology research; iv) alien species introduction causes a redistribution of species within ecosystems, but can be used to promote mitigation and conservation actions on native species or allow the development of new sampling strategies for large and deep lakes to obtain robust information on littoral occurrence of species; and v) researchers should focus their research also on cave organisms to solve the uncertainties linked to their poor taxonomic identification, sampling difficulty, biogeographic distribution, and richness contributing to solve Linnean, Wallacean, and Racovitzan shortfalls.

All the previous statements highlight that invertebrates are often neglected in biodiversity conservation policies that must be therefore implemented in order to contrast the current loss of biodiversity, to favor the achievement of the quality objectives of the Water Framework Directive, and to find mitigating solutions for the effects of anthropogenic pressures on aquatic ecosystems.

**Angela Boggero, Laura Garzoli**
*Editors*

 *water*

*Review*

# Assessing the Ecological Status of European Rivers and Lakes Using Benthic Invertebrate Communities: A Practical Catalogue of Metrics and Methods

Simon Vitecek [1,2,*], Richard K. Johnson [3] and Sandra Poikane [4]

1   Institute of Hydrobiology and Aquatic Ecosystem Management, University of Natural Resources and Life Sciences, 1180 Vienna, Austria
2   Wasser Cluster Lunz—Biologische Station GmbH, 3293 Lunz am See, Austria
3   Department of Aquatic Sciences and Assessment, Swedish University of Agricultural Sciences, 75007 Uppsala, Sweden; richard.johnson@slu.se
4   European Commission, Joint Research Centre (JRC), I-21027 Ispra, Italy; sandra.poikane@ec.europa.eu
*   Correspondence: simon.vitecek@boku.ac.at

**Abstract:** The Water Framework Directive requires that the ecological status of surface waters be monitored and managed if necessary. A central function in ecological status assessment has the Biological Quality Elements—organisms inhabiting surface waters—by indicating human impact on their habitat. For benthic invertebrates, a wide array of national methods are used, but to date no comprehensive summary of metrics and methods is available. In this study, we summarize the benthic invertebrate community metrics used in national systems to assess the ecological status of rivers, (very) large rivers, and lakes. Currently, benthic invertebrate assemblages are used in 26 national assessment systems for rivers, 13 assessment systems for very large rivers, and 21 assessment systems for lakes in the EU. In the majority of systems, the same metrics and modules are used. In the Red Queen's race of ecosystem management this may be a disadvantage as these same metrics and module likely depict the same stressors but there is growing evidence that aquatic ecosystems are subject to highly differentiated, complex multiple stressor impacts. Method development should be fostered to identify and rank impacts in multi-stressor environments. DNA-based biomonitoring 2.0 offers to detect stressors with greater accuracy—if new tools are calibrated.

**Keywords:** saprobic index; general degradation index; bioassessment



**Citation:** Vitecek, S.; Johnson, R.K.; Poikane, S. Assessing the Ecological Status of European Rivers and Lakes Using Benthic Invertebrate Communities: A Practical Catalogue of Metrics and Methods. *Water* **2021**, *13*, 346. https://doi.org/10.3390/w13030346

Academic Editors: Michele Mistri and Angela Boggero

Received: 5 December 2020
Accepted: 26 January 2021
Published: 30 January 2021

**Publisher's Note:** MDPI stays neutral with regard to jurisdictional claims in published maps and institutional affiliations.



## 1. Introduction

Protecting the integrity of the biodiversity and functioning of an ecosystem are key factors underpinning the continuous supply of ecosystem services [1]. In freshwater habitats, these are most importantly associated with supply of safe food and drinking water, self-purification, transportation, as well as recreation opportunities and are at the core of the Sustainable Development Goals [2,3].

The European Union implemented several laws to sustain natural resources and ensure environmental protection. As human impact on both biodiversity and ecosystem services increases such efforts are a primary concern of development and law-making [4–7]. The European initiatives and efforts are a good example how collaborative governing can help to overcome significant environmental challenges.

The Water Framework Directive (WFD, Directive 2000/60/EC) is among the most prominent pieces of legislation that pertain to EU freshwater and coastal habitats and prescribes monitoring the chemical (CS) and ecological status (ES) of surface waters—both lakes and rivers—in each EU member state. ES reflects the quality of the ecosystem structure and functioning of any surface water and is defined based on the deviation of observed communities of Biological Quality Elements (BQEs) from pristine or near-natural reference conditions. In particular, the river-specific assessment of ES is to be undertaken by assessing the composition and abundance of aquatic flora, or composition and abundance of benthic invertebrate fauna, or composition, abundance, and age structure of fish fauna. For lakes, assessment of ES also includes composition, abundance, and biomass of phytoplankton.

In line with the WFD, each EU member state implemented water body type-specific methods and tools to assess Ecological Status Class (ESC), and these approaches were intercalibrated to generate comparable results across the EU [8,9]. Assessing ESC of any water body follows a standard line of action [9–11]. In a first step, adequate and standardized sampling procedures are used to obtain a sample of the BQE community at a designated site. To obtain the relevant parameters of the BQE community, sampling focuses on measures composition, abundance, biomass, or age structure. Following this data generation step, specialized software solutions—hereafter called Ecological Status Class Assessment Tools (ESCATs)—are used to calculate values describing the community, and to relate these values to reference conditions and threshold values delimiting the different ESCs [8,9]. Based on deviation from reference and the threshold values in an ESC are assigned into five categories as: high, good, moderate, poor, or bad. A high ESC is defined as showing no to minimal deviations from a—theoretically pristine but in reality, mostly minimally disturbed—reference condition (sensu [12]), while a good ESC may reflect human activity but only to a slight extent. The other ESCs harbour communities that are significantly more disturbed than those observed in habitats of good ESC.

Naturally, a variety of different options were pursued to develop and ultimately intercalibrate ESC estimation tools, following monitoring traditions and available expertise [13,14]. However, a particularly prominent and frequently used BQE group is the benthic invertebrate fauna, and with excellent reason: benthic invertebrate assemblages are not only relatively easy to identify, but they also have narrow ecological niches which render them highly sensitive to changes in their environment—including anthropogenic disturbance [15–17]. Further, there is a strong tradition of using benthic invertebrates in biomonitoring, as their value as indicators of habitat conditions was recognized early (e.g., [18,19]). For benthic invertebrate assemblages, composition and abundance are to be measured and used for ESC assessment.

To quantify and compare these community parameters in ESCATs, different modules focusing on the sensitivity/tolerance and metrics are used. Modules are usually constructed based on taxon-specific indicator values or combinations of metrics that relate to the probability of a particular BQE community succession along a disturbance gradient. Based on composition and/or abundance of an observed BQE community all indicator values can then be summed or averaged, optionally including abundances as weights, to obtain a single numerical descriptor of the sampled habitat. Examples for modules include the Average Score Per Taxon index (ASPT), the Biological Monitoring Working Party index (BMWP), or the Saprobic index sensu Zelinka and Marvan [20–22]. Metrics usually are single numerical descriptors that are obtained by simple enumeration, via an alpha-diversity index (such as Margalef's index [23] or Shannon diversity [24]) or by calculating the proportion of a certain functional group observed in the BQE community (e.g., the number of sensitive or filter-feeding taxa) and are often used in combinations as multimetric indices [11,25].

However, there is surprisingly little information available on how ESCATs actually use benthic invertebrate assemblage data for WFD-compliant ESC estimation. In particular, there is a lack of comparative summaries for methods applied in rivers and lakes, and no attempt has yet been made to catalogue modules and metrics that are used in different ESCATs. Here, we provide a first summary of benthic invertebrate-based ESCATs used in rivers and lakes. We moreover present a catalogue of modules and metrics constituting ESCATs and discuss advantages and shortcomings of different modules and metrics for biomonitoring in general and specifically in respect to future biomonitoring approaches.

## 2. Data Acquisition and Access

Data on construction of national ESCATs were compiled from all primary articles (i.e., peer-reviewed articles and technical reports) that were submitted to the European Union in accordance with WFD regulations (see Supplementary Information), detailing metrics, indices, and modules used for ESC estimation in rivers, (very) large rivers, and lakes. Based on this database, we assessed which types of modules and metrics lay base to the respective assessment system. We did not include methods targeting acidification, as these were not implemented in each EU member state. Further, river-specific systems are used in (very) large rivers as well; this approach is shown separately for the purpose of this contribution. Further, different ESCATs are in use a number of countries, reflecting geographical differentiation.

We tabulated modules and metrics and assessed how frequently these are used in the diversity of ESCATs. For the purpose of this study, we consider as "modules" (usually used to refer to sensitivity/tolerance metrics [11]) tools that directly return an assessment result: an integrated index value from taxon-specific indicator values or metric combinations. Likewise, we consider as metrics numerical descriptors of bioindicator communities that deliver single values and can be integrated to a multimetric index. If a single module is used for assessments we treat it as depicting general degradation, as no further information on stressors is integrated. Based on our initial assessment, we developed a comparative framework in which the different national river ESCATs are grouped according to the number of shared modules and metrics.

## 3. Ecological Status Class Assessment across Europe

Currently, there are 26 assessment systems using benthic invertebrate assemblages for ESC estimation in rivers, 13 assessment systems for very large rivers—representing 38 ESCATS—and 21 assessment systems for lakes that represent 19 ESCATs.

For the assessment of rivers, three countries make use of decision tables: Denmark (Danish Stream Fauna Index, DSFI), and Bulgaria and Ireland (Q-value tables). Decision tables do not require computation of module or metric values, but rather assess ESC based on decision-table guided expert judgement. A total of six ESCATs are based on a single module only and used in Bulgaria (Q-value tables), Denmark, (Danish Stream Fauna Index, DSFI), Greece (Hellenic Evaluation Score, HES), Ireland (Q-value tables), Spain (Iberian BMWP), and Sweden (ASPT). All other ESCATs rely on the combination of at least one module and at least one metric. Of these, 18 are true multimetric ESCATs that integrate several metrics for ESC assessment (Table 1).

**Table 1.** Summary of national ESCATs for wadeable rivers, grouped according to similarities in modules and metrics used. Biomonitoring strategies differ among EU member states and associated countries, which is reflected in application of different modules and metrics. Bulgaria and Ireland use the Q-value approach, while Norway, Spain, Greece, Luxembourg, and Denmark use a single module. In the majority of EU member states, general degradation modules like the ASPT, the BMWP or the DSFI are complemented with additional metrics on diversity, functional ecology, and sensitivity/tolerance of benthic invertebrates. A large minority of ESCATs rely on a combination of organic pollution and general degradation modules with additional metrics. Abbreviations: O.P., organic pollution module; G.D., general degradation module; SI, Saprobic index; GDI, general degradation index; DI, diversity index; TD, taxonomic diversity metrics; CC, community composition metrics; FE, feeding ecology metrics; HM hydromorphology metrics; LC, life cycle metrics; ST, sensitive taxa metrics. All other abbreviations as listed in the glossary.

| Country | O.P. | G.D. | Taxonomic Diversity Metrics | | | Functional Metrics | | | Sens. Metrics |
|---|---|---|---|---|---|---|---|---|---|
| | SI | GDI | DI | TD | CC | FE | HM | LC | ST |
| BG | | Q-Value | | | | | | | |
| IE | | Q-Value | | | | | | | |
| NO | | GDI$_{ASPT}$ | | | | | | | |
| ES | | GDI$_{BMWP-I}$ | | | | | | | |

**Table 1.** *Cont.*

| Country | O.P. | G.D. | Taxonomic Diversity Metrics | | | Functional Metrics | | | Sens. Metrics |
| --- | --- | --- | --- | --- | --- | --- | --- | --- | --- |
| | SI | GDI | DI | TD | CC | FE | HM | LC | ST |
| EL | | $GDI_{EL}$ | | | | | | | |
| LU | | $GDI_{DSFI}$ | | | | | | | |
| DK | | $GDI_{DSFI}$ | | | | | | | |
| BE (W) | | $GDI_{GFI}$ | | $N_{Taxa}$ | | | | | |
| BE (F) | | $GDI_{BE}$ | $H'$ | $N_{Taxa}$, $N_{EPT}$ | | | | | $N_{Sens}$ |
| LV | | $GDI_{DSFI}$, $GDI_{ASPT}$ | | $N_{Taxa}$, $N_{EPT}$ | | | | | |
| LT | | $GDI_{DSFI}$, $GDI_{ASPT}$ | | $N_{DEP}$ | $P_{EHP}$, $P_{CrHi}$ | | | | |
| EE | | $GDI_{DSFI}$, $GDI_{ASPT}$ | $H'$ | $N_{Taxa}$, $N_{EPT}$ | | | | | |
| ES | | $GDI_{BMWP\text{-}I}$ | | $N_{Taxa}$, $N_{EPT}$, $N_{ETD}$ | $\log(P_{ETD})$, $\log(P_{EPTD})$ | | | | |
| ES | | $GDI_{ASPT\text{-}I}$ | | $N_{EPT}$, $N_{Fam}$ | $P_{EPTCD}$ | | | | |
| CY | | $GDI_{ASPT}$ | $H'$ | $N_{EPTFam}$, $N_{Fam}$, | $P_{EPTD}$, $P_{GOID}$ | | | | |
| IT | | $GDI_{ASPT}$ | $H'$ | $N_{Fam}$, $N_{EPT}$ | $P_{EPTD}$, $P_{GOID}$ | | | | |
| PL | | $GDI_{ASPT}$ | $H'$ | $N_{Fam}$, $N_{EPTFam}$ | $P_{EPTD}$, $P_{GOID}$ | | | | |
| PT | | $GDI_{ASPT\text{-}I}$ | $L'$ | $N_{Fam}$, $N_{EPTFam}$ | $P_{ETD}$, $P_{EPTCD}$ | | | | |
| FR | | $GDI_{ASPT}$ | $H'$ | $N_{Taxa}$ | | | | $P_{uvp}$, $P_{ovp}$ | |
| HU | | $GDI_{ASPT}$ | $H'$ | $N_{Taxa}$, $N_{EP}$, $N_{EPT}$, $N_{EPTCOB}$ | $P_{EPT}$ | $P_{Pre}$ | $P_{Rheo}$, $P_{Limno}$, $P_{Lit}$ | | |
| FI | | $GDI_{PMA}$ | | | | | | | $N_{Sens}$, $N_{Sens(EPT)}$ |
| NL | | | | | | | | | $P_{Sens}$, $P_{Pos}$, $P_{Neg}$ |
| ES | | | | $N_{Fam}$, $N_{EPTFam}$ | $P_{TP}$, $P_{DomFam}$, $P_{O}$; $\beta_{Bray\text{-}Curtis}$ | | | | $N_{Sens}$, $P_{Sens}$ |
| ES | | | $D'$ | | $P_{E}$, $P_{EPT}$; $\beta_{Bray\text{-}Curtis}$ | | | | $N_{Sens}$, $P_{Sens}$ |
| RO | $SI_{PB}$ | | $H'$ | $N_{Fam}$ | $P_{EPT}$, $P_{OCh}$ | $P_{Det}$ | $P_{Rheo}$, $P_{Limno}$ | | |
| SE | $SI_{ZM}$ | $GDI_{ASPT}$ | | $N_{EPT}$ | $P_{Cr}$, $P_{EPT}$ | | | | |
| SK | $SI_{ZM}$ | $GDI_{BMWP}$ | $D'$ | $N_{Fam}$, $N_{EPT}$ | | RETI, $P_{Det}$ | LZI, Rheo, $P_{MeR}$ | | |
| HR | $SI_{PB}$ | $GDI_{ASPT}$, $GDI_{HR}$ * | $D'$ | $N_{EPTCOB}$ | $P_{EPT}$ | RETI | LZI | $P_{r\text{-}Strat}$ | |
| SL | $SI_{ZM}$ | $GDI_{SL}$ * | $L'$ | $N_{EP}$ | $P_{EPT}$, $P_{T}$, $P_{P}$ | $P_{Det}$, $P_{XSAP}$ | | | |
| DE | $SI_{ZM}$ | $GDI_{GER}$ | | $N_{T}$, $N_{EPTCOB}$ | $P_{EPT}$ | | Rheo, $P_{Lit}$, $P_{Pel}$ | | |
| CZ | $SI_{ZM}$ | $GDI_{Bmodel}$ | $D'$ | $N_{Taxa}$, $N_{Chir}$ | $P_{EPT}$, $P_{E}$ | RETI | $P_{EpR}$, $P_{MeR}$, $P_{HyR}$, $P_{Lith}$ | | |
| AT | $SI_{ZM}$ | $GDI_{AUT}$ | $D'$ | $N_{Taxa}$, $N_{EPT}$ | $P_{EPT}$, $P_{OD}$ | RETI | $P_{Lit}$, LZI | | |

* depicts hydromorphology pressures as well.

In very large rivers, all existing methods integrate a saprobic and/or general degradation index with other metrics (Table 2).

**Table 2.** Summary of ESCATs used to assess ecological status specifically in (very) large rivers. Biomonitoring in very large rivers employs a similar range of metrics and modules as are used in wadeable rivers. All abbreviations and comments as in Table 1 and as listed in the glossary.

| Country | O.P. | G.D. | Taxonomic Diversity Metrics | | | Functional Metrics | | Sens. Metrics |
|---|---|---|---|---|---|---|---|---|
| | SI | GDI | DI | TD | CC | FE | HM | ST |
| AT | $SI_{ZM}$ | $GDI_{BMWP}$ | | | $P_{Ol}$ | RETI | LZI, $P_{ALP}$ | |
| BG | | | | $N_{Taxa}$ | | | | $N_{Sens}$ |
| HR | $SI_{PB}$ | $GDI_{HR}$ * | | | | | $P_{ALP}$ | |
| CZ | $SI_{ZM}$ | $GDI_{rekoMEPT}$ | $D'$ | | $P_{EPT}$ | | $P_{EpP}$ | $SPEAR_{organic}$ |
| EE | | $GDI_{ASPT}$ | $H'$ | $N_{Taxa}$, $N_{EPT}$ | | | | |
| DE | $SI_{ZM}$ | | | | | | PTI | |
| EL | | $GDI_{ASPT}$ | $H'$ | $N_{EPT}$, $N_{Fam}$ | $P_{EPTD}$, $P_{GOID}$ | | | |
| HU | | $GDI_{ASPT}$ | $H'$ | $N_{Taxa}$, $N_{EPTCOB}$ | | | | |
| LV | | $GDI_{ASPT}$ | $H'_{sensEPT}$ | $N_{Taxa}$, $N_{EPT}$ | | | | |
| NL | | | | $N_{EPTFam}$ | | | | $P_{Sens}$, $P_{Pos}$, $P_{Neg}$ |
| RO | $SI_{PB}$ | | $H'$ | $N_{Fam}$ | $P_{EPT}$, $P_{OCh}$ | | $P_{Limno}$ | |
| SK | $SI_{ZM}$ | $GDI_{BMWP}$ | | | $P_{Ol}$ | RETI | LZI, $P_{ALP}$ | |
| SL | $SI_{ZM}$ | $GDI_{SL}$ * | | | | | $P_{ALP}$ | |

For the assessment of lakes, two ESCATs are based on a single module only (used in Finland and Sweden, respectively), while the remainder of assessment approaches integrates at least one module and several metrics (Table 3).

**Table 3.** Summary of ESCATs used to assess ecological status of lakes in the European Union. Ecological status assessment in lakes focuses on modules and metrics detecting general degradation (GDI) as well as deviations in taxonomic diversity via diversity indices (DI), direct measurements of taxonomic diversity (TD), and community composition (CC). Further, functional metrics focusing on feeding ecology traits (FE) or habitat requirements concerning hydromorphology (HM) or life cycle traits (LC) of the observed benthic invertebrate assemblages are used. Additionally, sensitive taxa (ST) are used in ecological status assessment. All abbreviations as in Tables 1 and 2 and as listed in the glossary.

| Country | G.D. | Taxonomic Diversity Metrics | | | Functional Metrics | | | Sens. Metrics |
|---|---|---|---|---|---|---|---|---|
| | GDI | DI | TD | CC | FE | HM | LC | ST |
| AT | | | $N_{Taxa}$ | $LA_{Ol}$, $P_{Neo}$ | $P_{Gat}$ | | $P_{r/K\text{-strat}}$ | |
| BE | $GDI_{BE}$ | $H'$ | $N_{Taxa}$, $N_{EPT}$ | | | | | $N_{Sens}$ |
| BG | $GDI_{BMWP}$ | $H'$ | $N_{Fam}$ | | | | | |
| HR | | $D'$ | $N_{Fam}$ | $P_{EPT}$, $P_{Chi}$ | | | | |
| DK | $GDI_{ASPT}$ | $\ln(H')$ | $N_{EPTCOB}$ | $P_{COP}$ | | | | |
| EE | $GDI_{ASPT}$ | $H'$ | $N_{Taxa}$, $N_{EPT}$ | | | | | $ADI_{SE}$ |
| FI | $GDI_{PICM}$ # | | | | | | | |
| DE † | $GDI_{GER}$ | $H'$ | | $P_O$ | $P_{Gat}$ | | $P_{r/K\text{-strat}}$ | |
| DE † | $GDI_{GER}$ | | $N_{ETO}$ | $P_O$ | | $P_{Lith}$ | | |
| EL | | $H'$ | $N_{Taxa}$ | $P_{Chir}$ | | | | |
| EL | $GDI_{ASPT}$ | $D$ | | $P_O$ | | | | |
| HU | $GDI_{BMWP}$ | $H'$ | $N_{Fam}$ | | | | | |
| IT | $GDI_{BQIES}$ # | | | | | | | |
| LV | $GDI_{ASPT}$ | $H'$ | $N_{Taxa}$, $N_{EPTBO}$ | | | | | |
| LT | $GDI_{ASPT}$ | $\ln(H')$ | $N_{CEP}$ | $P_{COP}$ | | | | |
| NL | | | | | | | | $P_{Typ}$, $P_{Pos}$, $P_{Neg}$ |
| PL | $GDI_{ASPT}$ | $H'$ | $N_{EPTCOBFam}$ | $P_D$ | | | | |
| RO | | $H'$ | $N_{Fam}$ | $P_{ET}$, $P_G$, $P_{OrCh}$ | FGI | | | |
| SL | $GDI_{LFI}$ * | $D'$ | $N_{Taxa}$ | | | | | |
| ES | $GDI_{ABCO}$ | | $N_{CruIns}$ | | | | | |
| SE | $GDI_{SE}$ # | | | | | | | |

* addressing hydromorphological alterations; #, addressing eutrophication; †, Germany employs two different ESCATs for lakes, one for alpine and one for lowland lakes.

## 4. Types of Approaches: Decision Tables, Modules and Metrics

Decision tables present conditions that describe the status of an observed BQE community. These typically build on the occurrence and abundance of taxa and provide if-then solutions to assign ESC. Examples are the DSFI and the Q-value tables. The DSFI is also used outside of Denmark, in two other ESCATs (Estonia and Latvia).

Modules comprise Saprobic Indices (SI) and General Degradation Indices (GDI).

Metrics represent different aspects of the observed communities, such as captured by (1) a diversity index or a derivative thereof (e.g., H′, D′, L′, first Hill number), (2) raw taxonomic diversity (taxon numbers), (3) raw abundances (as density), (4) community composition metrics (proportions of taxa abundances), (5) metrics describing ecosystem function of the observed community (related mostly to feeding ecology or hydromorphological niches), (6) metrics quantifying sensitive taxa, (7) metrics describing phenology, (8) metrics describing reproductive strategies, and (9) metrics quantifying neozoa.

## 5. Catalogue of Modules and Metrics

### 5.1. Saprobic Indices

Saprobic indices (SI) were developed early on and are amongst the oldest approaches used to assess the status of aquatic ecosystems. They are based on the niche spaces occupied by different taxa, which can be expressed in ecological competence/preference points that serve as taxon-specific indicator values. These reflect the occurrence probability of indicator taxa along an ecological gradient of organic load, and, to a lesser degree, hydromorphology. Indicator values are available for a range of taxa, including not only benthic invertebrates but also aquatic flora. SIs are calibrated according to the ecological gradient observed in a specific region and describe the fit of the observed community to specific saprobic conditions; thus, various national adaptations of indicator values exist. SIs are currently used in seven national systems for rivers and large rivers each, but not in lakes. The most commonly used approaches were introduced by Pantle and Buck [18] (hereafter referred to as $SI_{PB}$) and Zelinka and Marvan [22] ($SI_{ZM}$). $SI_{PB}$ uses abundance of genus-level identified BQE in combination with taxon-specific indicator values to infer a saprobic index. $SI_{PB}$ is used in two river and two very large river ESCATs. By contrast, $SI_{ZM}$ relies on species-level identification and indicator weights in addition to indicator values for each taxon and integrates these values with the observed abundances to infer a saprobic index. $SI_{ZM}$ is used in six river and five very large river ESCATs. In both approaches, the observed SI is related to threshold values to infer a saprobic quality class or an ecological quality ratio based on saprobic conditions.

### 5.2. General Degradation Indices

General degradation indices (GDIs) follow the same principles as SIs, i.e., taxon-specific indicator values are developed based on occurrence of taxa along a disturbance gradient. The most commonly applied GDIs are the BMWP and the ASPT indices that rely on family-level identification of indicator taxa and require no abundance data [20,21]. This makes for a rapid and versatile application of these indices possible but comes with a trade-off concerning specificity and accuracy. Regionally specific GDIs were developed and calibrated to detect human-induced impairment with greater efficacy. National variants of river GDIs that comprise waterbody type-specific variants and threshold values exist in Austria, Belgium (Flanders), France, Germany, Greece, and Slovenia; adaptations of the French GDI are also used in other countries.

BMWP and ASPT: The Biological Monitoring Working Party and the Average Score Per Taxon indices are based on occurrence of families of benthic invertebrates. For each family, the assigned indicator value reflects occurrence probability in minimally disturbed or, ideally, pristine conditions. The BMWP is calculated as the sum of all indicator values and the ASPT is calculated as the BMWP value divided by the number of scoring (observed) families. The ASPT is used in 13 river, four very large river and six lake ESCATs, while the BMWP is used in three river, and two very large river and lake ESCATs each.

National GDIs: National GDIs can emulate the BMWP/ASPT approach but may also have higher specificity for particular water body types and may also include information on abundances of benthic invertebrates. Effectively, the various national GDIs follow the same principle in assigning indicator values to sets of taxa associated with specific habitat conditions, but usually rely on higher taxonomic resolution. GDIs are usually calibrated to detect impairment of habitats rather than specific stressors; an exception are the GDIs employed in Slovenia and Croatia, that specifically take hydromorphological alteration into account [26].

*5.3. Synopsis of Single Metrics Used as Benthic Invertebrate Community Descriptors*

Diversity indices are calibrated against reference conditions for use in biomonitoring. To this end, communities at sites along a disturbance gradient are sampled and their alpha diversity described by means of a diversity index. Margalef's index (D') [23,27], Shannon diversity (H') [24,27], and the corresponding Evenness (L') calculated as a derivative of Shannon diversity [27], are most commonly used. Alternatively, the First Hill Number calculated as the exponential function of Shannon diversity may be used. Calculation of diversity indices should be based on species-level identification and properly assessed abundances. Diversity indices follow different functions, according to their construction: Margalef's index follows a relatively linear function, while Shannon diversity follows a logarithmic function (but can be linearized by calculating its exponential function, as is done for the First Hill Number). For ESC estimation in rivers, Shannon diversity is most commonly used (7 ESCATs), followed by Margalef's index (4 ESCATs), and Evenness (2 ESCATs). Very large river methods may rely on Shannon diversity (4 ESCATs), Margalef's Index (1 ESCAT) or Shannon diversity computed based on a preselected set of sensitive EPT-taxa (1 ESCAT). In lakes, Shannon diversity is used in 10 ESCATs, Simpson's and Margalef's index in one ESCAT each and the First Hill Number in two ESCATs.

Raw taxa numbers (taxon richness) may be employed in addition to or as an alternative to diversity indices. Here, total diversity is expressed as number of taxa at a predefined taxonomic resolution encountered at a designated sampling site. Further, the number of taxa recorded in one or several groups can be used as metric. To this end large-bodied taxa are selected, and their diversity recorded at predefined taxonomic levels. Usually, number of Ephemeroptera, Plecoptera, Trichoptera, Diptera, Coleoptera, Bivalvia, Odonata, Oligochaeta, Gastropoda, or Chironomidae (a group of Diptera) taxa are employed, either singly or in combination. Combinations of taxa are constructed to indicate ecosystem integrity or impairment, and to make the metric more robust to changes in community composition. The number of bioindicator taxa encountered at a site usually reaches its peak in pristine or minimally disturbed conditions where microhabitat diversity and structure are unperturbed. Typically sets of taxa are summarized to obtain values, including the following metrics:

- Total taxa number ($N_{Taxa}$): Total number of taxa found in a sample; used in 18 river, six very large river and 10 lake ESCATs.
- Number of Diptera, Ephemeroptera, and Plecoptera taxa ($N_{DEP}$): used in a single river ESCAT.
- Number of Ephemeroptera and Plecoptera taxa ($N_{EP}$): used in two river ESCATs.
- Number of Ephemeroptera, Plecoptera, and Trichoptera taxa ($N_{EPT}$): used in 13 river, four very large river and two lake ESCATs.
- Number of Ephemeroptera, Plecoptera, Trichoptera, Coleoptera, Odonata, and Bivalvia taxa ($N_{EPTCOB}$): used in three river, one very large river and two lake ESCATs.
- Number of Trichoptera taxa ($N_T$): used in one river ESCAT.
- Number of Ephemeroptera, Trichoptera and Diptera taxa ($N_{ETD}$): used in one river ESCAT.
- Number of Ephemeroptera, Trichoptera, and Odonata taxa ($N_{ETO}$): used in one lake ESCAT.
- Number of Ephemeroptera, Plecoptera, Trichoptera, Bivalvia, and Odonata taxa ($N_{EPTBO}$): used in one lake ESCAT.

- Number of Coleoptera, Ephemeroptera, and Plecoptera taxa ($N_{CEP}$): used in one lake ESCAT.
- Number of Crustacea and Insecta taxa ($N_{CruIns}$): used in one lake ESCAT.
- Number of Chironomidae taxa ($N_{Chir}$): used in one river and one lake ESCAT.

Composition of the bioindicator community is an important metric, and usually focuses on sets of indicator taxa. These usually are the same relatively large-bodied taxa as targeted for a taxa-numbers metric, due to their relatively predictable occurrence in pristine/minimally disturbed or degraded conditions. In most cases community composition metrics are constructed taking abundances into account (either as raw abundances or abundance classes), so that proportions of indicator groups are compared. The response of this metric is a shift in proportions of taxa along a disturbance gradient and is assessed by calculating proportions that sets of taxa contribute to a particular benthic invertebrate taxa community.

- Proportion of Ephemeroptera, Plecoptera, and Trichoptera specimens ($P_{EPT}$): used in nine river, two very large river, and one lake ESCAT.
- Proportion of Oligochaeta and Diptera specimens ($P_{OD}$): used in one river ESCAT.
- Proportion of Oligochaeta and Chironomidae specimens ($P_{OCh}$): used in one river and very large river ESCAT each.
- Proportion of Oligochaeta specimens ($P_{Ol}$): used in one very large river and lake ESCAT each.
- Proportion of Ephemeroptera, Trichoptera, and Diptera specimens ($P_{ETD}$): used in two river ESCATs.
- Proportion of Ephemeroptera, Trichoptera, Plecoptera, and Diptera specimens ($P_{EPTD}$): used in four river and one very large river ESCATs.
- Proportion of Ephemeroptera, Heteroptera and Plecoptera specimens ($P_{EHP}$): used in one river ESCAT.
- Proportion of Crustacea and Hirudinea ($P_{CrHi}$): used in one river ESCAT.
- Proportion of Crustacea specimens ($P_{Cr}$): used in one river ESCAT.
- Proportions of Trichoptera and Plecoptera specimens ($P_{TP}$): used in one river ESCAT.
- Proportions of Ephemeroptera and Trichoptera specimens ($P_{ET}$): used in one lake ESCAT.
- Proportion of Ephemeroptera, Plecoptera, Trichoptera, Coleoptera, and Diptera specimens ($P_{EPTCD}$): used in two river ESCATs.
- Proportion of Diptera specimens ($P_{D}$): used in one river ESCAT.
- Proportion of Plecoptera specimens ($P_{P}$): used in one river ESCAT.
- Proportion of Trichoptera specimens ($P_{T}$): used in one river ESCAT.
- Proportion of Odonata specimens ($P_{O}$): used in three lake ESCAT.
- Proportion of Gastropoda ($P_{G}$): used in one lake ESCAT.
- Proportion of Orthocladiinae ($P_{OrCh}$): used in one lake ESCAT.
- Proportion of Chironomidae ($P_{Chir}$): used in one lake ESCAT.
- Proportion of Chironomini ($P_{Chi}$): used in one lake ESCAT.
- Proportion of Gastropoda, Oligochaeta, and Diptera specimens ($P_{GOID}$): used in three river ESCATs.
- Proportion of dominant taxa ($P_{Dom}$): used in one river ESCAT.
- Proportion of neozoa ($P_{Neo}$): used in one lake ESCAT.
- A beta-diversity index to quantify differences in community composition: the Bray-Curtis dissimilarity ($\beta_{Bray\text{-}Curtis}$) is used in two river ESCATs to characterize how well an observed benthic invertebrate fauna community and the reference community match.

Functional metrics based on benthic invertebrate assemblages are less frequently considered in EU assessments, and when used usually focus on feeding ecology or the hydromorphological niche of an observed community. Specific indices have been developed to describe integrated feeding guilds, but also simple proportions of single feeding types (e.g., predators) are used. Proportions of feeding guilds are assumed to follow a specific succession along the river continuum, and deviation therefrom can be quantified as signal of human impact. In particular, deviations in the proportions of feeding guilds may relate to organic input, increased sediment load, or changes in hydromorphology. Bioindicator communities likewise assemble according to natural hydromorphological gradients along rivers, where sections typically exhibit dominance of certain taxa. Quantifying changes in these dominance patterns can support identification of pressures related to hydromorphological alterations. In particular, functional metrics can be grouped in feeding guild metrics, hydromorphology metrics, life cycle metrics, and sensitive taxa metrics:

1.  Feeding guild metrics take prevalence of certain feeding strategies of benthic invertebrate assemblages into account and include:

    - Rhithron feeding type index (RETI) [28]: proportion of grazer and shredder taxa in the total share of specimens of grazer, shredder, filter-feeding and detritvorous taxa; used in four river and two very large river ESCATs.
    - Proportion of predators ($P_{Pre}$): share of specimens of predatory taxa; used in one river ESCAT.
    - Proportions of grazers ($P_{Gra}$): share of specimens of grazer taxa; used in one river ESCAT.
    - Proportions of detritivorous taxa ($P_{Det}$): Share of specimens of detritivorous taxa (feeding on detritus); used in one river and one lake ESCAT.
    - Proportions of gatherers ($P_{Gat}$): share of specimens of gathering taxa (feeding on benthic fine particulate organic matter); used in two lake ESCATs.
    - Proportion of xylal-feeding, shredder, active filter feeders and passive filter feeders ($P_{XSAP}$): share of specimens of xylal-feeding taxa (i.e., taxa feeding on wood), shredder taxa, active filter feeders (feeding on fine particulate organic matter that is actively filtered from the water body), and passive filter feeders (feeding on fine particulate organic matter that is passively filtered from the water body); used in one river ESCAT.

2.  Hydromorphological metrics assess prevalence of taxa occupying distinct hydromorphological niches and include the following metrics:

    - Rheo-Index (Rheo): share of rheophilic and rheobiont taxa in the total number of rheophilic, rheobiont, stagnophilic, stagnobiont, and ubiquitous taxa; used in two river ESCATs.
    - Longitudinal Zonation Index (LZI) [29]: analogous to the SI where calculation may follow Pantle and Buck [17] or Zelinka and Marvan [22]—describes the fit of the observed community to particular hydromorphological conditions by using taxon-specific ecological competence/preference points that describe the occurrence probability of a taxon along a hydromorphological gradient from spring to estuary; used in three river and two very large river ESCATs.
    - Potamon-Typie Index (PTI) [30]: describes how strongly an observed benthic invertebrate assemblage deviates from an expected near-natural or minimally disturbed state in large and very large rivers based on taxon-specific indicator values; used in one very large river ESCAT.
    - Proportion of littoral taxa ($P_{Lit}$): share of specimens of littoral associated taxa (i.e., taxa inhabiting the littoral zone of lakes); used in three river ESCATs.
    - Proportions of rheophilic taxa ($P_{rheo}$): share of specimens of rheophilic taxa (i.e., taxa associated with lotic habitats); used in two river ESCATs.
    - Proportions of limnophilic taxa ($P_{Limno}$): share of specimens of limnophilic taxa (i.e., taxa associated with lentic habitats); used in two river and 1 very large river ESCATs.
    - Proportion of epipotamal taxa ($P_{EpP}$): share of specimens of epipotamal-associated taxa; used in one very large river ESCAT.
    - Proportion of epirhithral taxa ($P_{EpR}$): share of specimens of epirhithral-associated taxa (i.e., taxa occurring predominately in epirhithral sections); used in one river ESCAT.

- Proportion of metarhithral taxa ($P_{MeR}$): share of specimens of metarhithral-associated taxa (i.e., taxa occurring predominately in metarhithral sections; used in two river ESCATs.
- Proportion of hyporhithral taxa ($P_{HyR}$): share of specimens of hyporhithral-associated taxa (i.e., taxa occurring predominately in hyporhithral sections); used in one river ESCAT.
- Proportion of akal-inhabiting, littoral and psammal-inhabiting taxa ($P_{ALP}$): share of specimens of akal-inhabiting, littoral or psammal-inhabiting taxa; used in one river and four very large river ESCATs.
- Proportion of lithal-inhabiting taxa ($P_{Lith}$): share of specimens of lithal-inhabiting taxa; used in one river and one lake ESCAT.
- Proportion of pelal-inhabiting taxa ($P_{Pel}$): Share of specimens of pelal-inhabiting taxa; used in one river ESCAT.

3. Life cycle metrics targeting phenology of mostly aquatic insect taxa focus on the proportion of univoltine (reproducing only once in a year) to polyvoltine (reproducing several times during a year) taxa. Additionally or alternatively, proportions of taxa following different reproductive strategies (r- or K-selected taxa) can provide information about the status of a waterbody. Metrics based on phenology and life cycle strategies include:

- Proportions of univoltine and polyvoltine taxa ($P_{upv}$): used in one river ESCAT.
- Proportion of ovoviviparous taxa ($P_{ovp}$): used in one river ESCAT.
- Proportion of r- and/or K-selected taxa ($P_{r\text{-}strat}$, $P_{r/K\text{-}strat}$): used one river and two lake ESCATs.

4. Sensitive taxa metrics are based on presence or abundance of sensitive taxa and rely on an a priori definition of sensitive taxa, according to the corresponding system and pressures. Here, presence/abundance of sensitive taxa can be calibrated to indicate impairment of an ecosystem. Either numbers of taxa or proportions of taxon abundances can be used. Metrics of this type are obtained by summing taxon richness in the respective categories or calculating proportions of sensitive taxa abundance and include:

- Number of sensitive taxa ($N_{Sens}$): used in four river, one very large river and one lake ESCAT.
- Number of sensitive Ephemeroptera, Plecoptera, and Trichoptera taxa ($N_{Sens(EPT)}$): used in one river ESCAT.
- Proportion of sensitive taxa ($P_{Sens}$): used in three river and one very large river ESCATs.
- Number of typical taxa ($N_{Typ}$): used in one river ESCAT.
- Proportion of typical taxa ($P_{Typ}$): used in one river ESCAT.
- Proportion of positive or negative taxa ($P_{Pos}$; $P_{Neg}$): used in one river, one very large river and one lake ESCAT each.
- Species-at-risk (SPEAR) by organic pollution: used in one very large river ESCAT.

## 6. Comparing ESCATs between Countries Based on the Most Frequently Used Modules and Metrics

In rivers, the most commonly used modules for organic pollution or general degradation comprise the $SI_{ZM}$, and the ASPT and BMWP. Among diversity indices, Shannon diversity is used most often in river ESCATs, followed by Margalef's index. Taxonomic diversity is usually assessed based on taxonomic richness and EPT richness metrics. Proportions of EPT taxa are also frequently used as a community composition metric. The most frequently used feeding guild metric is the RETI; the corresponding hydromorphological metrics are the LZI and $P_{Lit}$. Further, the number of sensitive taxa is commonly used in river ESCATs.

In (very) large rivers, a very similar set of modules and metrics are frequently used: the $SI_{ZM}$, the ASPT, Shannon diversity, overall taxonomic richness, and EPT richness as well as proportion of EPT taxa and the LZI. Additionally, proportions of Oligochaeta for community composition metrics and proportions of akal-, littoral-, and psammal-inhabiting taxa as hydromorphological metrics are used.

In lakes the most frequently used metrics differ slightly. Similar to river assessments, ASPT and BMWP are used as well as Shannon diversity, Margalef's Index, and taxonomic richness. However, lake assessments also frequently use proportions of Odonata, gatherers and may include proportions of r-selected and K-selected taxa.

In an attempt to generalize patterns of river ESCAT construction, we propose that, based on these patterns, four main groups can be distinguished: First, ESCATs relying exclusively on decision tables as used in Bulgaria, Ireland, Luxembourg, and Denmark. Second, ESCATs using a single module only as currently used in Norway, Spain, or Greece. Third, ESCATs relying on a combination of modules and metrics comprising at most the ASPT or a similar index, Shannon diversity, taxonomic richness and EPT richness and few if any other ecological metrics. ESCATs of the third group are used to assess river ecological status in Norway, Spain, Belgium (Wallonia), Belgium (Flanders), Latvia, Lithuania, Estonia, Cyprus, Italy, Poland, Portugal, and France. Fourth, ESCATs extensively using ecological metrics or pursuing altogether different strategies are used in ecological status assessments of rivers in Spain, Romania, Sweden, Slovakia, Croatia, Slovenia, Germany, the Czech Republic, and Austria. In this group, distinct indices were often developed to account for large ecological gradients represented by many different river types.

Concerning (very) large rivers and lakes, making coarse generalizations is challenging. Similarities of ESCATs exist, and in some cases (e.g., Austria and Slovenia) the same ESCATs are used, but the diversity of approaches developed for these systems as compared to river ESCATs is much greater.

## 7. Advantages of Different Module and Metric Types

The ways modules and metrics are defined follow different philosophies [10,11]. Modules and metrics can either be designed to allow for a rapid and robust assessment of an ecosystem, or to detect specific stressors that are particularly relevant for a habitat or set of habitats at high resolution to simplify management decisions.

For instance, fast and versatile application, reflected in coarse taxonomic resolution and limited integration of ecological parameters, may be favored over other more resource demanding approaches. Modules such as the ASPT or the BMWP are prime examples for this approach, as they do not require abundance data, are based on family-level taxonomic resolution, and can readily be applied to a broad spectrum of aquatic habitats [20,21,31]. Due to the ease of use, definition of ESC boundaries and establishing reference conditions can be speedily undertaken in typology-based approaches for estimating reference conditions. However, it should be noted that model-based estimates of reference conditions can outperform typology-based approaches if typology classification is not biologically meaningful [32–35]. Likewise, metrics based on taxon richness are robust and easily adopted [36]. Depending on the focal indicator taxa group, taxon richness may either decrease (e.g., total number of taxa, number of EPT taxa) or increase (e.g., number of Diptera taxa) with increasing stressor impact (e.g., [11,37,38]). However, they cannot be trained to a particularly high specificity when using coarse taxonomic resolution—if deviating from the reference benchmark their informativeness concerning the stressor is relatively limited [39–41]. Therefore such modules and metrics can be important tools when establishing ecological status of a hitherto unassessed habitat or when ESC estimation is to be conducted under resource-limited conditions. However, an ordination or modelling-based approach to a priori define reference conditions (and select metrics and modules) usually provides better resolution than the simple use of a taxonomically coarse index such as ASPT [42–46].

Diversity indices and community composition metrics take an intermediate position between rapid and high-resolution modules and metrics, requiring abundance or relative abundance data but not ecological information for metric calculation [44]. Both approaches quantify shifts in proportions of taxa under stressor impact that stem from differences in niche space occupied by individual taxa—resulting in clear deviance from reference conditions. In particular changes in the relative abundance of taxa associated with specific habitat conditions can be used to identify habitat modification. For instance, an increase in the relative abundance of Oligochaeta and Diptera may indicate an accumulation of fine sediments and organic matter at a sampling site [47]. Conversely, a decrease in the proportion of, e.g., Ephemeroptera, Plecoptera, and/or Trichoptera taxa may signal habitat homogenization (i.e., a man-made simplification of habitat conditions resulting in the loss of microhabitats), changes in food resource composition, or organic pollution [41,47–49].

Alternatively, modules and metrics may aim at resolving stressor impact at a high level of detail and focus on ecological characteristics of indicator species as well as abundance. As concerns modules, type-specific GDIs are especially useful for detecting and characterizing impairment on rivers but have not yet been established for lakes. Precisely calibrated GDI modules are highly relevant in many river ESCATs, and often are key in detecting stressor impact. Further, SIs are robust at quantifying the degree of anthropogenic organic load in rivers and can be calibrated to a high degree of specificity and accuracy [37]. Using SIs for ESC estimation in lakes is not as common, mostly because lake assemblages do not respond predictably to organic pollution (potentially due to a relatively greater proportion of benthic invertebrate taxa breathing atmospheric oxygen in these habitats) [17,50]. Other ecologically based modules have been developed following the SI example with substantial effort placed on acquiring the autecological characterization of species used as bioindicator taxa—culminating in a database now detailing ecological preferences all major bioindicator species [17]. The significance of such data for biomonitoring is tremendous: ecological metrics such as the RETI, or proportions of certain feeding guilds or taxa associated with specific hydromorphological conditions are widely used and enable differentiation of stressors [39,46,49]. In combination with properly defined reference conditions these high-resolution modules and metrics can be used to detect impact of organic pollution, hydromorphological alteration, or changes in land use relating to allochthonous matter input (e.g., large woody debris) [41,48,51,52]. In addition, metrics focusing on phenology of aquatic insects or reproductive strategies of the bioindicator communities can be used to assess long-term stability of an ecosystem and can be calibrated to detect impact of unrecorded disturbance events or the relatively slow response to climate change [53,54].

Ultimately, all of these different approaches have their advantages: either by providing rapid and easy assessment options or by providing precise information on the prevalent stressors. From a management perspective, both qualities are desirable and support decision-making. In light of the growing body of evidence for the complex interplay of multiple stressors in aquatic ecosystems [55–57], having precise information may, however, finally prove more important than getting that information quickly.

## 8. The Way Forward, Part 1: Improving ESC Assessment

To construct ESCATs, a combination of both rapid and high-resolution modules and metrics can be selected. Usually, however, only one of the two approaches is followed because of national assessment traditions and ambitions. A significant challenge for ESC assessment is the combination of multiple stressors, that all exert—in function of their combination and magnitude—distinct roles in different habitats [58,59]. In addition to this challenge, many of the currently used ESCATs lack information on stressor-response relationships, and thus may fail to identify stressors, or accurately rank stressor importance [59,60]. This is particularly true for rapidly applied modules and metrics with a long history of use and impedes designing and implementing best management measures. Improving ESC assessment will therefore require a shift towards modules and metrics based on ecological characteristics of the bioindicator communities as well as calibration of new and better ESCATs targeting the most important stressors [6,61,62]. Indeed, many of the currently used ESCATs were designed to depict impact of organic pollution—which, due to the implementation of the EU Directive on Urban Wastewater Treatment (91/271/EEC) and the WFD, in large areas of Europe no longer is the most pressing stressor—and do not cover emerging or multiple stressors including pollution by microplastics [6,61–64]. Naturally, assessment systems need to be adapted to reflect environmental changes brought about by the prevalent stressors. For aquatic ecologists, this is the Red Queen's race of ecosystem management: timely providing such tools as may serve to maintaining the integrity of ecosystems at a certain stage of societal development.

### 9. The Way Forward, Part 2: Development of Future Biomonitoring Tools

Improving ESC assessment can only be achieved by calibrating ESCATs to detect stressors and quantify the magnitude of their impact on aquatic ecosystems. An especially promising approach for this purpose is offered by the integration of modern molecular tools, such as DNA metabarcoding, in ESC assessment—effectively the implementation of biomonitoring 2.0 [65,66].

Implementing novel tools like DNA metabarcoding will require adapting novel ES-CATs. This is because DNA metabarcoding and other molecular techniques cannot deliver the exact same data on BQE communities as is currently used for assessment [67,68]. At present, for benthic invertebrates taxa lists at various levels of taxonomic identification are used in connection with (mostly) abundance data. The standard sampling and assessment protocols allow for establishing an area-standardized estimate of the taxon richness and individuals at a sampling site to produce a taxa x abundance matrix. Based on these data, community composition and abundance can be described and compared to reference conditions.

A generic biomonitoring 2.0 workflow can make use of samples obtained following these sampling protocols but may also be applied to environmental samples [69]. In the latter case, no voucher material is available for later quality control. Standard samples may be sorted to obtain the specimens, or preservative ethanol may be decanted and filtered to obtain material for DNA extraction. Following DNA extraction, PCR or a bait-capture approach may be used to enrich and subsequently sequence target gene fragments using high-throughput sequencing (HTS). Next, bioinformatic analyses deconstruct the sequencing raw data (usually containing several replicates of each sample) into molecular operational taxonomic units (MOTUs; groups of sequences derived by, e.g., a threshold-based approach that are treated as taxa) that can be assigned to true taxa by use of reference libraries. MOTUs assigned to the same true taxon can then be summarized, and the number of individual HTS reads combined to allow for an estimate of taxon-specific read numbers in the sample.

Throughout such a biomonitoring 2.0 workflow, critical and well-founded decisions must be made to adopt the most suitable molecular and bioinformatic methods to control potential sources of error, and to reliable and repeatable generate data for ESC assessment. Still, some limitations of molecular methods remain: molecular methods do not produce abundance data that is identical to that used in existing ESCATs, and often only deliver occurrence data with reasonably high plausibility. Due to stochastic and choice-induced processes, also taxa lists produced by molecular methods are not identical to those delivered by the currently used standard methods [70–73].

Acknowledging these differences between standard and molecular data, we expect that some modules and metrics may still be used in a biomonitoring 2.0 framework following re-calibration (i.e., re-definition of reference conditions using molecular data). This is particularly pertinent to taxa number metrics, and modules using occurrence data only such as various GDIs. Stringent re-definition of reference conditions and module and metric re-calibration will be necessary for other metrics, particularly for such as integrating ecological characteristics of bioindicator taxa in assessment. The list of the most frequently used modules and metrics presented here may serve as target to optimize performance of molecular tools for use in biomonitoring.

However, as at the same time ecological status class assessment is developed (e.g., [74]), the purpose of biomonitoring 2.0 should rather be to develop a comprehensive novel toolbox to win the Red Queen's race of ecosystem management instead of trying to follow in the same steps.

**Supplementary Materials:** The following are available online at https://www.mdpi.com/2073-4441/13/3/346/s1, Table S1: Supplementary information detailing references for national ESCATs and intercalibrated approaches as used for the assessment of ecological status according to the Water Framework Directive in lakes, rivers and very large rivers in the European Union and associated countries.

**Author Contributions:** Conceptualization, S.V. and S.P.; methodology, S.V.; formal analysis, S.V.; investigation, S.V. and S.P.; writing—original draft preparation, S.V.; writing—review and editing, S.V., R.K.J., and S.P. All authors have read and agreed to the published version of the manuscript.

**Funding:** This research received no external funding. Open access funding provided by BOKU Vienna Open Access Publishing Fund.

**Institutional Review Board Statement:** Not applicable.

**Informed Consent Statement:** Not applicable.

**Acknowledgments:** S.V. acknowledges the patience of his family (including an elderly dog) as this manuscript was entirely written while on paternity leave. We gratefully acknowledge the reviews of four academic referees whose comments improved this manuscript.

**Conflicts of Interest:** The authors declare no conflict of interest.

## Abbreviations

| | |
|---|---|
| $ADI_{SE}$ | Swedish Acidification Index |
| $\beta_{Bray\text{-}Curtis}$ | Bray–Curtis dissimilarity |
| $D$ | Simpson's Index |
| $D'$ | Margalef's Index |
| FGI | Functional Group Index |
| $GDI_{ABCO}$ | Spanish general degradation index for lakes |
| $GDI_{ASPT}$ | Average Score Per Taxon index |
| $GDI_{ASPT\text{-}I}$ | Iberian Average Score Per Taxon index |
| $GDI_{AUT}$ | Austrian general degradation index |
| $GDI_{BE}$ | Belgian (Flanders) general degradation index |
| $GDI_{Bmodel}$ | Czech general degradation index |
| $GDI_{BMWP}$ | Biological Monitoring Working Party index |
| $GDI_{BMWP\text{-}I}$ | Iberian Biological Monitoring Working Party index |
| $GDI_{BQIES}$ | Benthic Quality Index, Italian general degradation index |
| $GDI_{DSFI}$ | Danish Stream Fauna Index |
| $GDI_{EL}$ | Greek general degradation index |
| $GDI_{GER}$ | German general degradation index, German Fauna Index |
| $GDI_{GFI}$ | Groupe Faunistique Indicateurs, French/Belgian (Wallonie) general degradation index |
| $GDI_{HR}$ | Croatian general degradation index |
| $GDI_{LFI}$ | Lake Fauna Index, Slovenian general degradation index |
| $GDI_{PICM}$ | Profundal Invertebrate Community Metrics, Finnish general degradation index |
| $GDI_{PMA}$ | Percent Model Affinity, Finnish general degradation index |
| $GDI_{rekoMEPT}$ | Czech general degradation index for very large rivers |
| $GDI_{SE}$ | Swedish general degradation index (Dahl-Johnson index) |
| $GDI_{SL}$ | Slovenian general degradation index |
| $H'$ | Shannon diversity |
| $L'$ | Evenness; $LA_{Ol}$—Natural logarithm of Oligochaeta abundances |
| $\ln(H')$ | First Hill number |
| LZI | Longitudinal Zonation Index |
| $N_{CEP}$ | Number of Coleoptera, Ephemeroptera and Plecoptera taxa |
| $N_{Chir}$ | Number of Chironomidae taxa |

| | |
|---|---|
| $N_{CruIns}$ | Number of Crustacea and Insecta taxa |
| $N_{DEP}$ | Number of Diptera, Ephemeroptera and Plecoptera taxa |
| $N_{EP}$ | Number of Ephemeroptera and Plecoptera taxa |
| $N_{EPT}$ | Number of Ephemeroptera, Plecoptera and Trichoptera taxa |
| $N_{EPTCOB}$ | Number of Ephemeroptera, Plecoptera, Trichoptera, Coleoptera, Odonata and Bivalvia taxa |
| $N_{ETD}$ | Number of Ephemeroptera, Trichoptera and Diptera taxa |
| $N_{ETO}$ | Number of Ephemeroptera, Trichoptera and Odonata taxa |
| $N_{Sens}$ | Number of sensitive taxa |
| $N_{Sens(EPT)}$ | Number of sensitive EPT taxa |
| $N_T$ | Number of Trichoptera taxa |
| $N_{Taxa}$ | Total taxa number |
| $P_{ALP}$ | Proportion of akal-inhabiting, littoral and psammal-inhabiting taxa |
| $P_{Chi}$ | Proportion of Chironomini specimens |
| $P_{Chir}$ | Proportion of Chironomidae specimens |
| $P_{COP}$ | Proportion of Coleoptera, Odonata and Plecoptera specimens |
| $P_{Cr}$ | Proportion of Crustacea specimens |
| $P_{CrHi}$ | Proportion of Crustacea and Hirudinidae specimens |
| $P_D$ | Proportion of Diptera specimens |
| $P_{Det}$ | Proportion of detritivorous taxa (entails collectors and gatherers) |
| $P_{DomFam}$ | Proportion of specimens of dominant families |
| $P_E$ | Proportion of Ephemeroptera specimens |
| $P_{EHP}$ | Proportion of Ephemeroptera, Heteroptera and Plecoptera specimens |
| $P_{EpP}$ | Proportion of epipotamal-associated taxa |
| $P_{EpR}$ | Proportion of epirhithral-associated taxa |
| $P_{EPT}$ | Proportion of Ephemeroptera, Plecoptera and Trichoptera specimens |
| $P_{EPTCD}$ | Proportion of Ephemeroptera, Plecoptera, Trichoptera, Coleoptera and Diptera specimens |
| $P_{EPTD}$ | Proportion of Ephemeroptera, Plecoptera, Trichoptera and Diptera specimens |
| $P_{ET}$ | Proportion of Ephemeroptera and Trichoptera specimens |
| $P_{ETD}$ | Proportion of Ephemeroptera, Trichoptera and Diptera specimens |
| $P_G$ | Proportion of Gastropoda specimens |
| $P_{Gat}$ | Proportion of gathering taxa |
| $P_{GOlD}$ | Proportion of Gastropoda, Oligochaeta and Diptera specimens |
| $P_{HyR}$ | Proportion of hyporhithral-associated taxa |
| $P_{Limno}$ | Proportion of limnophilic taxa |
| $P_{Lit}$ | Proportion of littoral-associated taxa |
| $P_{Lith}$ | Proportion of lithal-associated taxa |
| $P_{MeR}$ | Proportion of metarhithral-associated taxa |
| $P_{Neg}$ | Proportion of «negative» taxa |
| $P_{Neo}$ | Proportion of Neozoa specimens |
| $P_O$ | Proportion of Odonata specimens |
| $P_{OCh}$ | Proportion of Oligochaeta and Chironomidae specimens |
| $P_{OD}$ | Proportion of Oligochaeta and Diptera specimens |
| $P_{Ol}$ | Proportion of Oligochaeta specimens |
| $P_{OrCh}$ | Proportion of Orthocladiinae (Chironomidae) specimens |
| $P_{ovp}$ | Proportions of ovoviviparous taxa |
| $P_P$ | Proportion of Plecoptera specimens |
| $P_{Pel}$ | Proportions of pelal-inhabiting taxa |
| $P_{Pos}$ | Proportions of «positive» taxa |
| $P_{Pre}$ | Proportions of predators |
| $P_{r\text{-}strat}$ | Proportions of r-selected taxa |
| $P_{r/K\text{-}strat}$ | Proportions of r- and K-selected taxa |
| $P_{Rheo}$ | Proportions of rheophilic taxa |
| $P_{Sens}$ | Proportions of sensitive taxa |
| $P_T$ | Proportion of Trichoptera specimens |

| PTI | Potamon-Typie Index |
| $P_{TP}$ | Proportion of Trichoptera and Plecoptera specimens |
| $P_{Typ}$ | Proportions of typical taxa |
| $P_{upv}$ | Proportions of uni- and polyvoltine taxa |
| $P_{XSAP}$ | Proportions of xylal-feeding, shredder, active filter feeders and passive filter feeders |
| Q-Value | Q-Value tables |
| RETI | Retention Feeding Type Index |
| Rheo | Rheo Index |
| $SI_{PB}$ | Saprobic Index sensu Pantle and Buck |
| $SI_{ZM}$ | Saprobic Index sensu Zelinka and Marvan |
| $SPEAR_{organic}$ | Species-at-risk by organic pollution |

## References


1. Grizzetti, B.; Liquete, C.; Pistocchi, A.; Vigiak, O.; Zulian, G.; Bouraoui, F.; De Roo, A.; Cardoso, A.C. Relationship between ecological condition and ecosystem services in European rivers, lakes and coastal waters. *Sci. Total Environ.* **2019**, *671*, 452–465. [CrossRef] [PubMed]

2. Boulton, A.J.; Ekebom, J.; Gislason, G.M. Integrating ecosystem services into conservation strategies for freshwater and marine habitats: A review. *Aquat. Conserv.* **2016**, *26*, 963–985. [CrossRef]

3. Griggs, D.; Stafford-Smith, M.; Gaffney, O.; Rockström, J.; Öhman, M.C.; Shyamsundar, P.; Steffen, W.; Glaser, G.; Kanie, N.; Noble, I. Sustainable development goals for people and planet. *Nature* **2013**, *495*, 305–307. [CrossRef] [PubMed]

4. Dodds, W.K.; Perkin, J.S.; Gerken, J.E. Human impact on freshwater ecosystem services: A global perspective. *Environ. Sci. Technol.* **2013**, *47*, 9061–9068. [CrossRef]

5. Dudgeon, D. Multiple threats imperil freshwater biodiversity in the Anthropocene. *Curr. Biol.* **2019**, *29*, 960–967. [CrossRef]

6. Reid, A.J.; Carlson, A.K.; Creed, I.F.; Eliason, E.J.; Gell, P.A.; Johnson, P.T.J.; Kidd, K.A.; MacCormack, T.J.; Olden, J.D.; Cooke, S.J.; et al. Emerging threats and persistent conservation challenges for freshwater biodiversity. *Biol. Rev.* **2019**, *94*, 849–873. [CrossRef]

7. Culhane, F.; Teixeira, H.; Nogueira, A.J.A.; Borgwardt, F.; Trauner, D.; Lillebø, A.; Piet, G.; Kuemmerlen, M.; McDonald, H.; Robinson, L.A.; et al. Risk to the supply of ecosystem services across aquatic ecosystems. *Sci. Total Environ.* **2019**, *660*, 611–621. [CrossRef]

8. Birk, S.; Willby, N.J.; Kelly, M.G.; Bonne, W.; Borja, A.; Poikane, S.; Van de Bund, W. Intercalibrating classifications of ecological status: Europe's quest for common management objectives for aquatic ecosystems. *Sci. Total Environ.* **2013**, *454*, 490–499. [CrossRef]

9. Poikane, S.; Birk, S.; Böhmer, J.; Carvalho, L.; de Hoyos, C.; Gassner, H.; Hellsten, S.; Kelly, M.; Solheim, A.L.; van de Bund, W.; et al. A hitchhiker's guide to European lake ecological assessment and intercalibration. *Ecol. Indic.* **2015**, *52*, 533–544. [CrossRef]

10. Birk, S.; Bonne, W.; Borja, A.; Brucet, A.; Courrat, A.; Poikane, S.; Solimini, A.; Van De Bund, W.; Zampoukas, N.; Hering, D. Three hundred ways to assess Europe's surface waters: An almost complete overview of biological methods to implement the Water Framework Directive. *Ecol. Indic.* **2013**, *18*, 31–41. [CrossRef]

11. Hering, D.; Feld, C.K.; Moog, O.; Ofenböck, T. Cook book for the development of a Multimetric Index for biological condition of aquatic ecosystems: Experiences from the European AQEM and STAR projects and related initiatives. In *The Ecological Status of European Rivers: Evaluation and Intercalibration of Assessment Methods*; Furse, M.T., Hering, D., Brabec, K., Buffagni, A., Sandin, L., Verdonschot, P., Eds.; Springer: Dordtrecht, The Netherlands, 2006; pp. 311–324.

12. Stoddard, J.L.; Larsen, D.P.; Hawkins, C.P.; Johnson, R.K.; Norris, R.H. Setting expectations for the ecological condition of streams: The concept of reference condition. *Ecol. Appl.* **2006**, *16*, 1267–1276. [CrossRef]

13. Hering, D.; Buffagni, A.; Moog, O.; Sandin, L.; Sommerhäuser, M.; Stubauer, I.; Feld, C.; Johnson, R.; Pinto, P.; Zahrádková, S.; et al. The Development of a System to Assess the Ecological Quality of Streams Based on Macroinvertebrates–Design of the Sampling Programme within the AQEM Project. *Int. Rev. Hydobiol.* **2003**, *88*, 345–361. [CrossRef]

14. Hering, D.; Moog, O.; Sandin, L.; Verdonschot, P.F.M. Overview and application of the AQEM assessment system. *Hydrobiologia* **2004**, *516*, 1–20. [CrossRef]

15. Hering, D.; Johnson, R.K.; Kramm, S.; Schmutz, S.; Szoszkiewicz, K.; Verdonschot, P.F.M. Assessment of European streams with diatoms, macrophytes, macroinvertebrates and fish: A comparative metric-based analysis of organism response to stress. *Freshw. Biol.* **2006**, *51*, 1757–1785. [CrossRef]

16. Johnson, R.K.; Hering, D.; Furse, M.T.; Clarke, R.T. Detection of ecological change using multiple organism groups: Metrics and uncertainty. *Hydrobiologia* **2006**, *566*, 115–137. [CrossRef]

17. Schmidt-Kloiber, A.; Hering, D. www. freshwaterecology. info–An online tool that unifies, standardises and codifies more than 20,000 European freshwater organisms and their ecological preferences. *Ecol. Indic.* **2015**, *53*, 271–282. [CrossRef]

18. Pantle, R.; Buck, H. Die biologische Überwachung der Gewässer und die Darstellung der Ergebnisse. *Gas und Wasserfach* **1955**, *96*, 604.

19. Beer, W.D. Methodologische Untersuchungen zur Biologischen Fliess-Gewässeranalyse. *Internat. Rev. Hydrobiol. Und Hydrogr.* **1961**, *46*, 5–17. [CrossRef]

20. Armitage, P.D.; Moss, D.; Wright, J.F.; Furse, M.T. The performance of a new biological water quality score system based on macroinvertebrates over a wide range of unpolluted running-water sites. *Water Res.* **1983**, *17*, 333–347. [CrossRef]

21. Hawkes, H.A. Origin and development of the biological monitoring working party score system. *Water Res.* **1998**, *32*, 964–968. [CrossRef]

22. Zelinka, M.; Marvan, P. Zur Präzisierung der biologischen Klassifikation der Reinheit fließender Gewässer. *Arch. Hydrobiol.* **1961**, *57*, 389–407.

23. Margalef, R. Diversidad de Especies en las Comunidades Naturales. *Publ. Inst. Biol. Apl.* **1951**, *9*, 5–27.

24. Shannon, C.E. A mathematical theory of communication. *Bell Syst. Tech. J.* **1948**, *27*, 379–423. [CrossRef]

25. Karr, J.R.; Chu, E.W. *Biological Monitoring and Assessment: Using Multimetric Indexes Effectively*; University of Seattle: Washington, WA, USA, 1997; pp. 1–155.

26. Urbanić, G. Hydromorphological degradation impact on benthic invertebrates in large rivers in Slovenia. *Hydrobiologia* **2014**, *729*, 191–207. [CrossRef]

27. Magurran, A.E.; McGill, B.J. *Biological Diversity: Frontiers in Measurement and Assessment*; Oxford University Press: New York, NY, USA, 2011; pp. 1–359.

28. Schweder, H. Rhithron-Ernährungstypen-Index (RETI)–ein Parameter zur Beschreibung und Bewertung der Ernährungsbeziehungen von Makroinvertebraten in kleinen Fliessgewässern. In *Erweiterte Zusammenfassung DGL-Jahrestagung 1990*; Deutsche Gesellschaft für Limnologie: Essen, Germany, 1990; pp. 325–329.

29. Moog, O. *Fauna Aquatica Austriaca*; Wasserwirtschaftskataster, Bundesministerium für Land-und Forstwirtschaft: Wien, Austria, 1995.

30. Schöll, F.; Haybach, A. Bewertung von großen Fließgewässern mittels Potamon-Typie-In- dex (PTI). Verfahrensbeschreibung und Anwendungsbeispiele. *Bundesanst. für Gewässerkunde Mitt.* **2001**, *23*, 1–28.

31. Birk, S.; Hering, D. Direct comparison of assessment methods using benthic macroinvertebrates: A contribution to the EU Water Framework Directive intercalibration exercise. In *The Ecological Status of European Rivers: Evaluation and Intercalibration of Assessment Methods*; Furse, M.T., Hering, D., Brabec, K., Buffagni, A., Sandin, L., Verdonschot, P., Eds.; Springer: Dordtrecht, The Netherlands, 2006; pp. 401–415.

32. Pardo, I.; Gómez-Rodríguez, C.; Wasson, J.-G.; Owen, R.; van de Bund, W.; Kelly, M.; Bennett, C.; Birk, S.; Buffagni, A.; Ofenböck, G.; et al. The European reference condition concept: A scientific and technical approach to identify minimally-impacted river ecosystems. *Sci. Total Environ.* **2012**, *420*, 33–42. [CrossRef]

33. Davy-Bowker, J.; Clarke, R.T.; Johnson, R.K.; Kokes, J.; Murphy, J.F.; Zahradkova, S. A comparison of the European Water Framework Directive physical typology and RIVPACS-type models as alternative methods of establishing reference conditions for benthic macroinvertebrates. In *The Ecological Status of European Rivers: Evaluation and Intercalibration of Assessment Methods*; Furse, M.T., Hering, D., Brabec, K., Buffagni, A., Sandin, L., Verdonschot, P., Eds.; Springer: Dordtrecht, The Netherlands, 2006; pp. 91–105.

34. Aroviita, J.; Koskenniemi, E.; Kotanen, J.; Hämäläinen, H. A priori typology-based prediction of benthic macroinvertebrate fauna for ecological classification of rivers. *Environ. Manage.* **2008**, *42*, 894–906. [CrossRef]

35. Johnson, R.K.; Hallstan, S. Modelling outperforms typologies for establishing reference conditions of boreal lake and stream invertebrate assemblages. *Ecol. Indic.* **2018**, *93*, 864–873. [CrossRef]

36. Johnson, R.K.; Wiederholm, T.; Rosenberg, D.M. Freshwater biomonitoring using individual organisms, populations, and species assemblages of benthic macroinvertebrates. In *Freshwater biomonitoring and benthic macroinvertebrates*; Rosenberg, D.M., Resh, V., Eds.; Springer: New York, NY, USA, 1993; pp. 40–158.

37. Ofenböck, T.; Moog, O.; Gerritsen, J.; Barbour, M. A stressor specific multimetric approach for monitoring running waters in Austria using benthic macro-invertebrates. *Hydrobiologia* **2004**, *516*, 251–268. [CrossRef]

38. Sánchez-Montoya, M.M.; Vidal-Abarca, M.R.; Suárez, M.L. Comparing the sensitivity of diverse macroinvertebrate metrics to a multiple stressor gradient in Mediterranean streams and its influence on the assessment of ecological status. *Ecol. Indic.* **2010**, *10*, 896–904. [CrossRef]

39. Sandin, L.; Hering, D. Comparing macroinvertebrate indices to detect organic pollution across Europe: A contribution to the EC Water Framework Directive intercalibration. In *Integrated Assessment of Running Waters in Europe*; Hering, D., Verdonschot, P.F.M., Moog, O., Sandin, L., Eds.; Springer: Dordecht, The Netherlands, 2004; pp. 55–68.

40. Schmidt-Kloiber, A.; Nijboer, R.C. The effect of taxonomic resolution on the assessment of ecological water quality classes. *Hydrobiologia* **2004**, *516*, 269–283. [CrossRef]

41. Sundermann, A.; Gerhardt, M.; Kappes, H.; Haase, P. Stressor prioritisation in riverine ecosystems: Which environmental factors shape benthic invertebrate assemblage metrics? *Ecol. Indic.* **2013**, *27*, 83–96. [CrossRef]

42. Jyväsjärvi, J.; Aroviita, J.; Hämäläinen, H. An extended Benthic Quality Index for assessment of lake profundal macroinvertebrates: Addition of indicator taxa by multivariate ordination and weighted averaging. *Freshwater Sci.* **2014**, *33*, 995–1007. [CrossRef]

43. Reynoldson, T.B.; Rosenberg, D.M.; Resh, V.H. Comparison of models predicting invertebrate assemblages for biomonitoring in the Fraser River catchment, British Columbia. *Can. J. Fish. Aquat. Sci.* **2001**, *58*, 1395–1410. [CrossRef]

44. Lakew, A.; Moog, O. A multimetric index based on benthic macroinvertebrates for assessing the ecological status of streams and rivers in central and southeast highlands of Ethiopia. *Hydrobiologia* **2015**, *751*, 229–242. [CrossRef]

45. Hering, D.; Meier, C.; Rawer-Jost, C.; Feld, C.K.; Biss, R.; Zenker, A.; Sundermann, A.; Lohse, S.; Böhmer, J. Assessing streams in Germany with benthic invertebrates: Selection of candidate metrics. *Limnologica* **2004**, *34*, 398–415. [CrossRef]

46. Buckland, S.T.; Magurran, A.E.; Green, R.E.; Fewster, R.M. Monitoring change in biodiversity through composite indices. *Philos. Trans. R Soc. B* **2005**, *360*, 243–254. [CrossRef]

47. Jones, J.I.; Murphy, J.F.; Collins, A.L.; Sear, D.A.; Naden, P.S.; Armitage, P.D. The impact of fine sediment on macro-invertebrates. *River Res. Appl.* **2012**, *28*, 1055–1071. [CrossRef]

48. Townsend, C.R.; Uhlmann, S.S.; Matthaei, C.D. Individual and combined responses of stream ecosystems to multiple stressors. *J. Appl. Ecol.* **2008**, *45*, 1810–1819. [CrossRef]

49. Gieswein, A.; Hering, D.; Feld, C.K. Additive effects prevail: The response of biota to multiple stressors in an intensively monitored watershed. *Sci. Total Environ.* **2017**, *593*, 27–35. [CrossRef]

50. O'Toole, C.; Donohue, I.; Moe, S.J.; Irvine, K. Nutrient optima and tolerances of benthic invertebrates, the effects of taxonomic resolution and testing of selected metrics in lakes using an extensive European data base. *Aquat. Ecol.* **2008**, *42*, 277–291. [CrossRef]

51. Marzin, A.; Archaimbault, V.; Belliard, J.; Chauvin, C.; Delmas, F.; Pont, D. Ecological assessment of running waters: Do macrophytes, macroinvertebrates, diatoms and fish show similar responses to human pressures? *Ecol. Indic.* **2012**, *23*, 56–65. [CrossRef]

52. Mellado-Díaz, A.; Sánchez-González, J.R.; Guareschi, S.; Magdaleno, F.; Toro Velasco, M. Exploring longitudinal trends and recovery gradients in macroinvertebrate communities and biomonitoring tools along regulated rivers. *Sci. Total Environ.* **2019**, *695*, 133774. [CrossRef] [PubMed]

53. Mondy, C.P.; Usseglio-Polatera, P. Using conditional tree forests and life history traits to assess specific risks of stream degradation under multiple pressure scenario. *Sci. Total Environ.* **2013**, *461*, 750–760. [CrossRef] [PubMed]

54. Conti, L.; Schmidt-Kloiber, A.; Grenouillet, G.; Graf, W. A trait-based approach to assess the vulnerability of European aquatic insects to climate change. *Hydrobiologia* **2014**, *721*, 297–315. [CrossRef]

55. Jackson, M.C.; Loewen, C.J.G.; Vinebrooke, R.D.; Chimimba, C.T. Net effects of multiple stressors in freshwater ecosystems: A meta-analysis. *Glob. Change Biol.* **2016**, *22*, 180–189. [CrossRef]

56. Johnson, R.K.; Angeler, D.G.; Hallstan, S.; Sandin, L.; McKie, B.G. Decomposing multiple pressure effects on invertebrate assemblages of boreal streams. *Ecol. Indic.* **2017**, *77*, 293–303. [CrossRef]

57. Johnson, R.K.; Hallstan, S.; Zhao, X. Disentangling the response of lake littoral invertebrate assemblages to multiple pressures. *Ecol. Indic.* **2019**, *85*, 1149–1157. [CrossRef]

58. Birk, S.; Chapman, D.; Carvalho, L.; Spears, B.M.; Andersen, H.E.; Argillier, C.; Auer, S.; Baattrup-Pedersen, A.; Banin, L.; Hering, D.; et al. Impacts of multiple stressors on freshwater biota across spatial scales and ecosystems. *Nat. Ecol. Evol.* **2020**, *4*, 1060–1068. [CrossRef]

59. Poikane, S.; Herrero, F.S.; Kelly, M.G.; Borja, A.; Birk, S.; van de Bund, W. European aquatic ecological assessment methods: A critical review of their sensitivity to key pressures. *Sci. Total Environ.* **2020**, *740*, 140075. [CrossRef]

60. Lemm, J.U.; Feld, C.K.; Birk, S. Diagnosing the causes of river deterioration using stressor-specific metrics. *Sci. Total Environ.* **2019**, *651*, 1105–1113. [CrossRef] [PubMed]

61. Carvalho, L.; Mackay, E.B.; Cardoso, A.C.; Baattrup-Pedersen, A.; Birk, S.; Blackstock, K.L.; Borics, G.; Borja, A.; Feld, C.K.; Solheim, A.L.; et al. Protecting and restoring Europe's waters: An analysis of the future development needs of the Water Framework Directive. *Sci. Total Environ.* **2019**, *658*, 1228–1238. [CrossRef] [PubMed]

62. Cid, N.; Bonada, N.; Heino, J.; Cañedo-Argüelles, M.; Crabot, J.; Sarremejane, R.; Soininen, J.; Stubbington, R.; Datry, T. A Metacommunity Approach to Improve Biological Assessments in Highly Dynamic Freshwater Ecosystems. *BioScience* **2020**, *70*, 427–438. [CrossRef] [PubMed]

63. Vigiak, O.; Grizzetti, B.; Udias-Moinelo, A.; Zanni, M.; Dorati, C.; Bouraoui, F.; Pistocchi, A. Predicting biochemical oxygen demand in European freshwater bodies. *Sci. Total Environ.* **2019**, *666*, 1089–1105. [CrossRef] [PubMed]

64. Gallitelli, L.; Cesarini, G.; Cera, A.; Sighicelli, M.; Lecce, F.; Menegoni, P.; Scalici, M. Transport and Deposition of Microplastics and Mesoplastics along the River Course: A Case Study of a Small River in Central Italy. *Hydrology* **2020**, *7*, 90. [CrossRef]

65. Baird, D.J.; Hajibabaei, M. Biomonitoring 2.0: A new paradigm in ecosystem assessment made possible by next-generation DNA sequencing. *Mol. Ecol.* **2012**, *21*, 2039–2044. [CrossRef]

66. Pawlowski, J.; Kelly-Quinn, M.; Altermatt, F.; Apothéloz-Perret-Gentil, L.; Beja, P.; Boggero, A.; Borja, A.; Bouchez, A.; Cordier, T.; Kahlert, M.; et al. The future of biotic indices in the ecogenomic era: Integrating (e)DNA metabarcoding in biological assessment of aquatic ecosystems. *Sci. Total Environ.* **2018**, *637–638*, 1295–1310. [CrossRef]

67. Erdozain, M.; Thompson, D.G.; Porter, T.M.; Kidd, K.A.; Kreutzweiser, D.P.; Sibley, P.K.; Swystun, T.; Chartrand, D.; Hajibabaei, M. Metabarcoding of storage ethanol vs. conventional morphometric identification in relation to the use of stream macroinvertebrates as ecological indicators in forest management. *Ecol. Indic.* **2019**, *101*, 173–184. [CrossRef]

68. Elbrecht, V.; Leese, F. Can DNA-based ecosystem assessments quantify species abundance? Testing primer bias and biomass-sequence relationships with an innovative metabarcoding protocol. *PLoS ONE* **2015**, *10*, e0130324. [CrossRef]

69. Mächler, E.; Little, C.J.; Wüthrich, R.; Alther, R.; Fronhofer, E.A.; Gounand, I.; Harvey, E.; Hürlemann, S.; Walser, J.-C.; Altermatt, F. Assessing different components of diversity across a river network using eDNA. *Environ. DNA* **2019**, *1*, 290–301. [CrossRef]

70. Gleason, J.E.; Elbrecht, V.; Braukmann, T.W.A.; Hanner, R.H.; Cottenie, K. Assessment of stream macroinvertebrate communities with eDNA is not congruent with tissue-based metabarcoding. *Mol. Ecol.* **2020**. [CrossRef] [PubMed]

71. Beentjes, K.K.; Speksnijder, A.G.C.L.; Schilthuizen, M.; Hoogeveen, M.; Pastoor, R.; van der Hoorn, B. Increased performance of DNA metabarcoding of macroinvertebrates by taxonomic sorting. *PLoS ONE* **2019**, *14*, e0226527. [CrossRef] [PubMed]

72.    Meyer, A.; Boyer, F.; Valentini, A.; Bonin, A.; Ficetola, G.F.; Beisel, J.J.N.; Bouquerel, J.; Wagner, P.; Gaboriaud, C.; Leese, F.; et al. Morphological vs. DNA metabarcoding approaches for the evaluation of stream ecological status with benthic invertebrates: Testing different combinations of markers and strategies of data filtering. *Mol. Ecol.* **2020**. [CrossRef] [PubMed]
73.    Bush, A.; Compson, Z.G.; Monk, W.A.; Porter, T.M.; Steeves, R.; Emilson, E.; Gagne, N.; Hajibabaei, M.; Roy, M.; Baird, D.J. Studying Ecosystems With DNA Metabarcoding: Lessons From Biomonitoring of Aquatic Macroinvertebrates. *Front. Ecol. Evol.* **2019**, *7*, 434. [CrossRef]
74.    Leitner, P.; Borgwardt, F.; Birk, S.; Graf, W. Multiple stressor effects on benthic macroinvertebrates in very large European rivers–a typology-based evaluation of faunal responses as a basis for future bioassessment. *Sci. Total Environ.* **2020**, 143472. [CrossRef]

 *water*

*Article*

# Biological, Chemical, and Ecotoxicological Assessments Using Benthos Provide Different and Complementary Measures of Lake Ecological Status

**Roberta Bettinetti** [1,2,*], **Silvia Zaupa** [2], **Diego Fontaneto** [2] **and Angela Boggero** [2]

1   Department of Human Sciences and Innovation for the Territory, University of Insubria, DiSUIT, Via Valleggio 11, 22100 Como, Italy
2   National Research Council, Water Research Institute (CNR-IRSA), Corso Tonolli 50, 28922 Verbania, Italy; silvia.zaupa@irsa.cnr.it (S.Z.); diego.fontaneto@irsa.cnr.it (D.F.); angela.boggero@irsa.cnr.it (A.B.)
*   Correspondence: roberta.bettinetti@uninsubria.it

Received: 12 March 2020; Accepted: 14 April 2020; Published: 16 April 2020

**Abstract:** The Water Framework Directive (WFD) aims to monitor continental water bodies in Europe to achieve good ecological status. Indexes based on biological quality elements (BQEs), ecotoxicological tests, and chemical characterizations are commonly used with standardized protocols to assess sediment quality and the associated risks. Here, we compare the results of quality assessment of benthic macroinvertebrates as BQEs as required by the WFD with the results of ecotoxicological tests and assessment of selected persistent organic pollutants (POPs) in sediments of the same eight water bodies in Italy. The aim was to verify if the assessment of quality through macroinvertebrates through POPs analyses and ecotoxicological tools can yield comparable, overlapping, or complementary results. We used the Benthic Quality Index (BQIES) for macroinvertebrates (two different applications), legacy POPs (dichloro-diphenyl-trichloroethane and metabolites (DDTs) and polychlorinated-biphenyls (PCBs)), and the emergence ratio (ER) and development rate (DR) for ecotoxicology. The results showed that the two indices within each approach were highly correlated, but between approaches, each result can lead to a completely different scenario, with rather different results of the assessment of ecosystem quality. The most striking result was that very few significant correlations existed between sediment quality assessment through macroinvertebrates and the risk assessment through analyses of micropollutants and ecotoxicological tests. The highest absolute r-value (0.81) was for the correlation between the BQIES$_{bottom}$ index and PCBs for micropollutants, whereas all other pairwise comparisons between indices had r-values ranging between 0.07 and 0.53. Our analysis calls for a caveat in the blind application of one or only a few indices of water/sediment quality, as the results of a single index may not represent the complexity of a freshwater ecosystem.

**Keywords:** biological quality element; chemical analysis; *Chironomus riparius*; DDTs; legacy contaminants; PCBs; POP; standard ecotoxicological tests; Water Framework Directive

---

## 1. Introduction

The Water Framework Directive (WFD—Directive 2000/60/EC) of the European Community has the aim of achieving a good qualitative ecological status for all types of water bodies in Europe. The importance of this piece of legislation is that water is considered as an exhaustible resource to be protected, emphasizing the role of aquatic ecology in management decisions. Several implementations of monitoring and assessment methods across Europe have been developed [1–4]. The evaluation of water quality within the WFD requires an integrated approach based on the use of specific biological metrics and on analyses of chemical compounds and hydro-morphological conditions [5].

Regarding biological indices, the composition and characteristics of the living communities of European waters are used as a primary focus to assess the quality of lakes and rivers as a whole. Several biological metrics using biological quality elements (BQEs) were set up by different European countries to address different risks [1,6–9], trying to harmonize classification systems across Europe. Regarding lakes, in Italy, the Benthic Quality Index (BQIES) was developed and implemented considering eutrophication as the main pressure [10,11], because of its importance in the national territory, given that nearly 41% of the Italian lakes are eutrophic [12]. The BQIES is based on detailed taxonomical identification on macroinvertebrates, mostly at the species level [13].

Regarding the chemical assessment of lake sediments within the WFD, this has to be performed with legally binding Environmental Quality Standards (EQSs) for selected chemical pollutants, known as priority substances, of EU-wide concern [14]. These include persistent organic pollutants (POPs) such as dichloro-diphenyl-trichloroethane and metabolites (DDTs) and polychlorinated-biphenyls (PCBs). A well-known chemical constrain can be the fact that not all potentially dangerous substances can be assessed, such as PCBs for which there are no EQSs yet. Moreover, chemical analyses of these substances are not always straightforward considering that contaminants are often present in the environment in mixtures [15].

An additional environmental test that is not yet included in the WFD involves the use of laboratory bioassays that involve a direct combination of chemistry (i.e., pollution) and biology (i.e., effect on living organisms). Laboratory ecotoxicology bioassays could evaluate the potential risk of toxic substances in the benthic environment, even without identifying the substances themselves [16], bypassing the impediment of the current implementation of the WFD, which does not include all potentially dangerous substances. These bioassays involve direct toxicity assays of environmental samples that are transferred to the laboratory and analyzed for toxicity against selected organisms [17]. Such samples contain the combination of the different pollutants present *in situ* and enable factors such as the bioavailability of contaminants and their interactions (synergistic and antagonistic) to be simultaneously studied (see [18] for more details).

The aim of the present paper was to compare the results of quality assessment of lakes obtained using different methods. Analyses of macroinvertebrates and analyses of organic pollutants according to the Italian standards of the WFD were compared with the results of security assessments obtained through ecotoxicological bioassays in the laboratory using sampled sediments according to standard guidelines with larvae of the chironomid midge *Chironomus riparius*. The purpose of the present study was to assess whether the three approaches (biological, chemical, ecotoxicological) provide similar and overlapping results, or whether each of them will provide different results for different facets of water quality and security.

## 2. Materials and Methods

### 2.1. Study Area

In all, eight natural and exploited lakes were selected as representative of the following aspects.

- Different weather and climate conditions, belonging to the Alpine (AL) and the Mediterranean (ME) Ecoregions covering the typical Italian climate conditions.
- Different types according to the Italian classification system (WFD requests) (Table 1 [19]): six of the eight lakes belong to the Alps and are in North-Western Italy (Piedmont region), the other two lakes belong to the insular Italy (Sardinia region) (Figure S1); the analyzed lakes cover five groups according to the national classification system in relation to their abiotic characteristics of altitude, surface area, mean depth, and catchment geo-lithology.
- Different origin, with five lakes being natural (mainly of glacial origin) and three representing the results of an artificial impoundment (reservoirs-Table S1). Also, different water uses are represented: the alpine reservoir is used as hydro-power generation plant, and the Mediterranean reservoirs as drinking water supplies.

- Different trophic level covering a gradient from ultra-oligotrophic to eutrophic conditions.

Thus, given their differences, any consistent pattern when comparing the lakes can provide supported information for reliable inference on the comparison between approaches in water quality and security assessment.

### 2.2. Sampling Methodology

We followed the Italian national protocol for monitoring macroinvertebrates for the application of BQIES [20]. In detail, water bodies were sampled in spring and autumn along transects connecting littoral, sub-littoral, and bottom areas of each lake. For micropollutants and ecotoxicological analyses, only the bottom area, where pollutants accumulate, was sampled and considered for subsequent analyses.

For biological analysis, an area of 675 $cm^2$ of sediment was sampled with a grab. The collected samples were sieved in the field through a 250-μm mesh net, fixed with 5% formalin, and bottled. Samples for micropollutants and ecotoxicological analyses were collected through the same grab. In this case, all samples were brought to the laboratory for subsequent analyses and immediately frozen for preservation.

### 2.3. Biological Assessment

In the lab, samples were sorted under a stereomicroscope, and specimens were separated into main groups, identified to the species level, when possible, and counted. The identification manuals were those in use at national and international level [21–23]. The Benthic Quality Index (BQIES) [24] was applied to each sampling station to evaluate the lake ecological status. Spatial (littoral, sub-littoral, and bottom) and temporal (spring and autumn) replicated samples were averaged to obtain a mean annual value representing the whole lake ecological status classification according to WFD requests. The BQIES is calculated based on the Shannon Diversity Index (SDI) corrected for the sensitivity values (indicator weights) attributed to each species; it can take values between 0, indicating low biological quality of the lake, and 1, indicating high quality near to the reference conditions [24]. Values of BQIES were averaged for the whole lake assessment ($BQIES_{all}$), or used as a separate value for the bottom samples only ($BQIES_{bottom}$) in different analyses, to allow a more direct comparison with micropollutants and ecotoxicological indices.

### 2.4. POPs Assessment

For POP determination, extraction was performed from 1 g of the freeze-dried and homogenized sediments in glass microfiber thimbles (19 mm internal diameter × 90 mm external length, Whatman, Maidstone, England) for 2 h with 60 mL of n-hexane (Carlo Erba, Cornaredo, Italy, pesticide analysis grade) using a modified Soxhlet apparatus (Velp Scientifica - ECO 6 thermoreactor).

Each organochlorine compound was recovered by several n-hexane washings and the extracts were concentrated down to approximately 2 mL and passed through a Florisil column (4 cm × 0.7 cm) with HCl-activated Cu powder (0.1 g) on the top. The Florisil column was eluted with 25 mL of n-hexane-dichloromethane (Carlo Erba, Cornaredo, Italy, pesticide analysis grade) using an 85:15 (v/v) mixture, and the eluate was concentrated to exactly 0.5 mL. The purified extracts were analyzed by gas-chromatography (GC Carlo Erba, Top 8000) coupled with a 63Ni electron capture detector - ECD 80 (Carlo Erba, Cornaredo, Italy), heated at 320 °C, using an on-column injection system (volume injected: 1 μL). The column was a Wall Coated Open Tubular Column fused silica CP-Sil-8 CB (50 m × 0.25 mm, film thickness: 0.25 μm, Varian, Harbor city, CA, USA). The temperature program used was as follows: from 60 °C to 180 °C at 20 °C·$min^{-1}$, followed by a run from 180 °C to 200 °C at 1.5 °C·$min^{-1}$; a further run was implemented from 200 °C to 270 °C at 3 °C·$min^{-1}$, followed by a final isothermal maintenance at 270 °C for 20 min, with helium as carrier gas (1 mL·$min^{-1}$) and nitrogen as auxiliary gas (30 mL·$min^{-1}$).

Quantification of DDT included 1,1-dichloro-2,2-bis (p-chlorophenyl) ethylene (pp′DDE), 1,1-dichloro-2,2-bis (p-chlorophenyl) ethane (pp′DDD), and 1,1,1-trichloro-2,2-bis (p-chlorophenyl) ethane (pp′DDT) as congeners of DDT and was performed using the external reference standards pp′DDE, pp′DDD, and pp′DDT (Pestanal, Sigma-Aldrich, Germany) in iso-octane (Carlo Erba, Italy, pesticide analysis grade). Quantification of 21 PCBs was performed on PCB 28 + 31, PCB 52, PCB 95, PCB 101, PCB 149 + 118, PCB 153, PCB 138, PCB 170, PCB 174, PCB 177; PCB 180; PCB 183, PCB 187, PCB 194, PCB 201, PCB 203, PCB 195, PCB 206, PCB 209.

The detection and quantification limits of the method varied from 0.05 to 0.1 $ng \cdot g^{-1}$ d.w. (dry weight) for DDTs and from 0.1 to 0.5 $ng \cdot g^{-1}$ d.w. for PCBs, depending on the organochlorine compound. The recovery efficiency was tested on a reference sediment previously used in an intercalibration exercise [25], and it was found to be within 80–100% for the DDTs and approximately 90% for each PCB congener. POPs data are presented as concentration per dry weight of the sediments ($ng \cdot g^{-1}$ d.w.). For the assessment, we used two summary metrics: total concentration of the analyzed DDT and total concentration of the analyzed PCB.

### 2.5. Ecotoxicological Assessment

The test organism *Chironomus riparius* Meigen 1804, of a strain maintained in the lab of the University of Insubria, was bred at 21 ± 1 °C under daily photoperiod in 40 L aquaria with control sediment (3 cm deep) as substrate. An 8 cm-deep column of dechlorinated tap water (hardness: 320 mg/L $CaCO_3$) was maintained over the sediment. The cultures were fed weekly with 1 g TetraMin fish food per tank, and the water was almost completely renewed every 2 months.

Bioassays were performed according to guideline 218 of the Organization for Economic Cooperation and Development (OECD) [18]. One day before the addition of first-instar larvae, 250 mL glass beakers were filled with collected sediments that were previously sieved (500 μm); 3.5 mL of a 4 $g \cdot L^{-1}$ water suspension of fish food, corresponding to 14 mg d.w. TetraMin, was put into each beaker. The contents of the beakers were allowed to settle in the dark at 21 ± 1 °C for 24 h. Five replicated beakers were prepared for each site, including the control (prepared following [18]). At the start of the test, 10 first-instar larvae chosen at random were transferred to each beaker. Tests were performed under a 16:8 h light:dark photoperiod for 10 days, with constant aeration. Every 3 days, the larvae were fed with 3.5 mL TetraMin suspension, and the evaporated water was added. Dissolved oxygen, pH and $NH^{4+}$ were measured in all the beakers at $t_0$ (beginning of the test) and at the end of the test ($t_{end}$) to verify standardized test conditions (Table S1). Every day all emerged adults were counted. The exposure lasted at maximum 28 days.

The bioassay measured the total number of animals that emerged and their sex. More animals are expected to emerge in samples with lower concentration of pollutants. The sum of midges emerged per vessel was determined and divided by the number of larvae introduced (emergence ratio (ER)) and the mean time span between the introduction of larvae (day 0 of the test) and the emergence of the experimental cohort of midges (development rate (DR)) was calculated. ANOVA (ANalysis Of Variance) and Dunnet *post hoc* test [26] were used to assess differences between treatments and control.

### 2.6. Statistical Analyses

The main question was to assess whether and to what extent biological indices, micropollutant indices, and ecotoxicological indices correlated between each other. To address this question, we compared the biological indices for macroinvertebrates (BQIES) applied both to the entire lake and only to the bottom area with the results of organic micropollutants (EQSs) for both total DDT and total PCB, and from DR and ER bioassays on *Chironomus riparius*. BQIES values for each lake were obtained by averaging the values for the different depths and different sampling seasons in the case of whole lake assessment ($BQIES_{all}$); BQIES values were also obtained only for bottom areas for assessment at maximum depth ($BQIES_{bottom}$); for micropollutants and ecotoxicological assessment, we used the data from the bottom samples. We performed the comparison using simple Spearman's rank correlation tests [26].

## 3. Results

### 3.1. Biological Assessment

Macroinvertebrates were represented by 12,483 individuals, divided in 136 taxa at the level of species, genus, or family depending on the phylum (Annelida, Arthropoda, Mollusca, and Plathyhelminthes) due to the presence of young-of-the-year organisms (see SM1 associated with [13]).

Oligochaetes and chironomids represented the most abundant groups, whose relative abundances together constituted from 46% to 100% of each macroinvertebrate lake sampling station (Figure 1).

The BQIES index provided a whole lake assessment of the ecological status with values varying between 0.52 in Lake Mergozzo and 0.22 in Lake Sirio (Table 1). Five lakes out of eight revealed values below the 0.4-threshold fixed by Italian regulations to separate good and moderate status. The BQIES applied only to the bottom samples showed values in the range between 0.004 (Lake Viverone) and 0.439 (Morasco reservoir); only the latter had good ecological status even in the deepest areas, whereas the remaining lakes showed moderate (Posada reservoir) or bad bottom ecological status (the other six lakes).

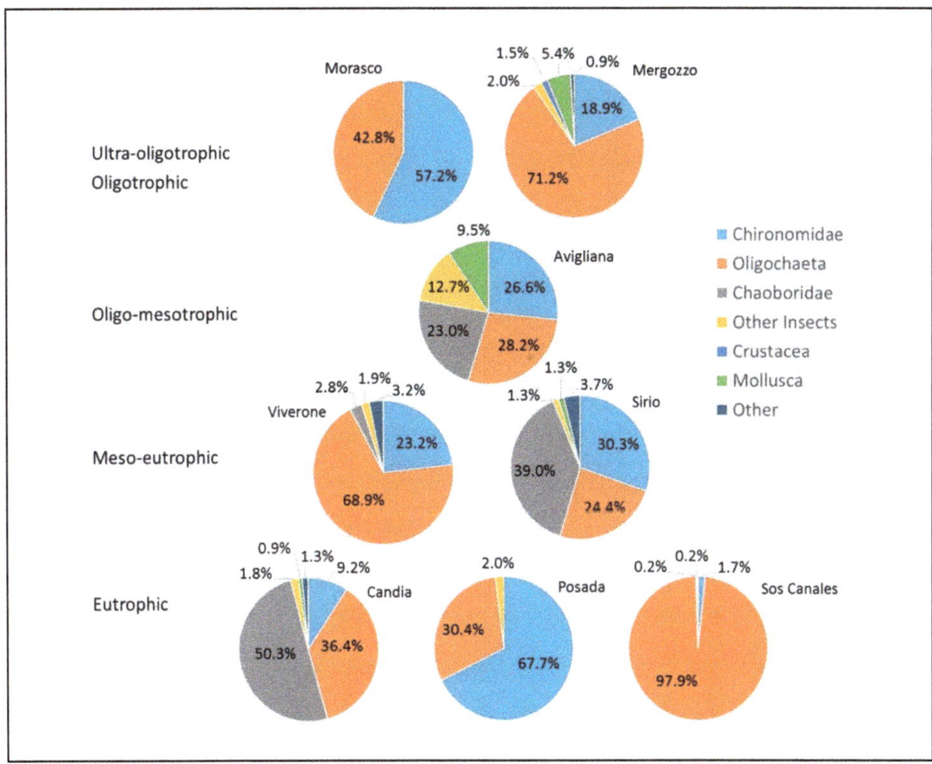


**Figure 1.** Annual mean macroinvertebrate assemblages (expressed as percentages) in each study lake. Trophic conditions of each lake are highlighted.

### 3.2. Micropollutants Assessment

Total DDT concentrations ranged between 0.7 ng·g$^{-1}$ d.w. in Morasco reservoir and 62.8 ng·g$^{-1}$ d.w. in Lake Sirio; total PCB concentrations ranged between 2.2 ng·g$^{-1}$ d.w. in Morasco reservoir and 61.5 in Lake Sirio (Table 1).

**Table 1.** Summary of each metric assessing water quality in the eight analysed lakes. The Benthic Quality Index for whole-lake assessment (BQIES$_{all}$) and for bottom lake assessment (BQIES$_{bottom}$) are adimensional numbers; DDTs and PCBs are expressed as ng·g$^{-1}$ d.w.; DR refers to the development ratio of chironomids; ER refers to the emergence ratio of chironomids. DDTs refers to dichloro-diphenyl-trichloroethane and metabolites, PCBs refers to polychlorinated-biphenyls; NA: not available.

| Lake Name | BQIES$_{all}$ | BQIES$_{bottom}$ | DDTs | PCBs | DR | ER |
|---|---|---|---|---|---|---|
| Avigliana piccolo | 0.26 | 0.01 | 7.27 | 19.4 | 0.051 | 55 |
| Candia | 0.42 | 0.21 | 14.73 | 29.8 | NA | NA |
| Sirio | 0.22 | 0.01 | 62.84 | 61.5 | 0.052 | 50 |
| Viverone | 0.26 | 0.00 | 25.44 | 38.1 | 0.051 | 70 |
| Mergozzo | 0.52 | 0.14 | 40.15 | 31.9 | 0.041 | 53 |
| Morasco | 0.46 | 0.44 | 0.75 | 2.2 | 0.053 | 51 |
| Posada | 0.37 | 0.37 | 10.36 | 2.7 | 0.049 | 70 |
| Sos Canales | 0.33 | 0.20 | 8.10 | 6.9 | NA | NA |

### 3.3. Ecotoxicological Assessment

Dissolved oxygen, pH and NH$^{4+}$ in water remained quite constant during the expositions (Table S2). All the values were within the range of acceptability of the OECD method. Only in some cases, a decrease of pH was observed with values <6. The sediments of Lake Candia and Sos Canales reservoir were very acidic at the end of the exposure. Oxygen was always higher than 60%. Controls were within the OECD guidelines. Controls showed always absence of NH$^{4+}$ both at the beginning and at the end of the test. Lakes Candia and Mergozzo, and Sos Canales and Posada reservoirs had a final NH$^{4+}$ concentration of 7.74 mg·L$^{-1}$, while the other lakes had always NH$^{4+}$ concentrations below 0.8 mg·L$^{-1}$.

Chironomids started to emerge after 17 days of exposure. Emergence in controls was within the validity criterion of OECD guidelines (ER > 70%); a minor ER in lake sediments (with the exception of Posada reservoir) occurred (Figure S2), and a significant difference (ANOVA with Dunnett *post hoc* test: $p < 0.05$) was found between all lakes and the controls. In the case of Lake Candia and Sos Canales reservoir no comparison was possible because chironomids died, probably due to the acidity of the medium. DR was significantly higher in Morasco reservoir, while Posada reservoir and Lake Mergozzo showed a minor DR indicating a specific toxicity for both lakes.

### 3.4. Comparison Between Assessments

The most striking result was that very few significant correlations existed between sediment quality assessment through macroinvertebrates, and risk assessment through micropollutants, and ecotoxicological tests (Figure 2). The highest absolute r-value (0.81) was between the BQIES$_{bottom}$ index and PCBs for micropollutants, whereas all other pairwise comparisons between indices had r-values ranging between 0.07 and 0.53.

Correlation values between indices within the same approach were high for micropollutants: DDTs and PCBs had a significant r-value of 0.88 (Figure 2). BQIES$_{all}$ (whole-lake assessment) and BQIES$_{bottom}$ (only bottom assessment) revealed a not significant but still high correlation (0.68; Figure 2). The correlation of DR and ER within ecotoxicological tests had an absolute r-value of 0.54 and was not significant (Figure 2).

Within each approach, our results confirm that using only one index, either DDTs or PCBs for micropollutants, and either BQIES$_{all}$ or BQIES$_{bottom}$ for macroinvertebrates, could be considered a reliable choice, because the two indices are significantly correlated. The same cannot be stated for DR and ER from ecotoxicological bioassays, potentially also due to the failure in two out of eight cases.

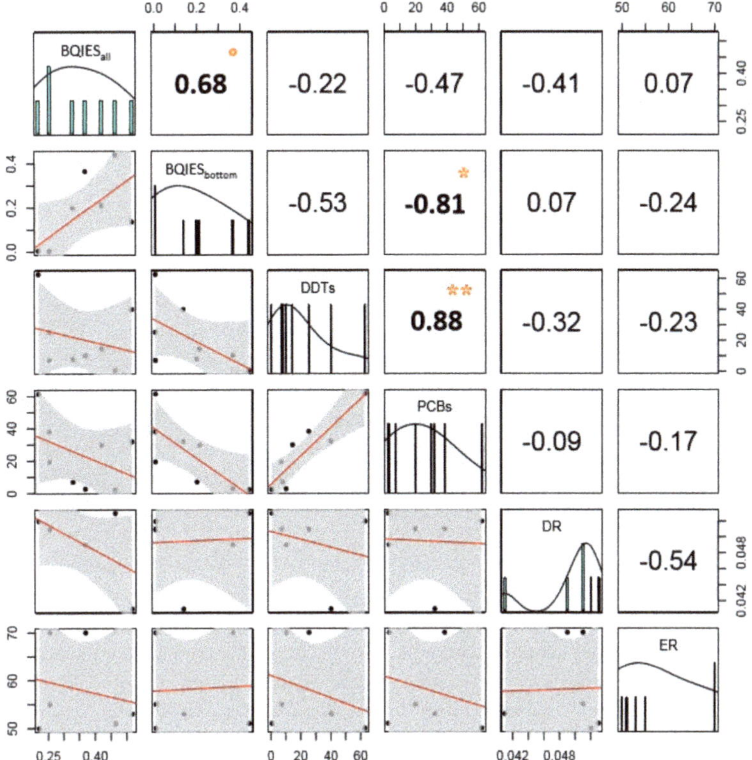

**Figure 2.** Pairwise correlations between the six analysed indices of water quality and security: BQIES$_{all}$ and BQIES$_{bottom}$ for ecological assessment, DDTs and PCBs for micropollutants, and DR and ER for ecotoxicological assessment. The cells in the diagonal show the histogram of distribution and the density plot for each index (see Table 1 for raw data); the cells below the diagonal show the scatterplots and the estimated linear correlation (colored line) with 95% confidence interval (grey areas) for each pairwise correlation; the cells above the diagonal report Spearman's r value of each pairwise correlation, with bold values indicating significant correlations and asterisks according to the $p$-value significance level (° for $p < 0.10$, * for $p < 0.05$, ** for $p < 0.01$). DR refers to the development ratio of chironomids; ER refers to the emergence ratio of chironomids. DDTs refers to dichloro-diphenyl-trichloroethane and metabolites, PCBs refers to polychlorinated-biphenyls 4.

## 4. Discussion

The main message of our study is that biological, chemical, and ecotoxicological approaches provide complementary and non-redundant information for a holistic assessment of ecological quality in lakes, when considering sediments as a model system. This seems a rather obvious result: different indices based on completely different elements should not provide the same answer to the question of quality and security assessment. Yet, what could be considered obvious may pass unnoticed to decision-makers and public administrators, given that sediment quality assessment is evaluated using only one or few indices simultaneously [7], and that water quality and risk assessment become pivotal in matching water demand and supply (domestic, industry, and agriculture supplies, but also for recreational, aesthetic, and nature conservation purposes).

Differences between biological, chemical, and ecotoxicological approaches to assess sediment quality and security are indeed expected [27]. For example, even if PCBs and DDTs are known to affect the test species (*Chironomus riparius*) we used for the ecotoxicological approach, the effects seen on the species may be also due to other contaminants acting in synergy. Nowadays, no reference values exist to predict the effects of contaminants in sediments on organisms [28].

An implementation of this aspect comes from the Environmental Quality Standards (EQS) approach for water, sediment, and biota, developed and implemented under the WFD. EQS are set for annual average concentrations (AA-EQS) and/or maximum admissible concentrations (MAC-EQS). In 2013, a new European Directive, 2013/39/EC, amended the Directives 2000/60/EC and 2008/105/EC regarding priority substances in the field of water policy, adding newly identified substances and revising the EQS of some existing substances. Following this new approach, a characterization with indexes of sediment quality seems still achievable.

Usually, a threshold effect concentration (TEC) has been used. Such TEC for total PCBs when no effects are observed is 40 ng·g$^{-1}$ d.w.; in our case, this concentration was never exceeded with the exception of Lake Sirio, where the total PCBs concentration was 60.55 ng·g$^{-1}$ d.w. (Table 1). In a study on Lake Maggiore contamination [29], the lowest observed effect concentration (LOEC) for total DDT in sediments on chironomids was 80.5 ng·g$^{-1}$, and no samples taken during this research showed similar concentrations. It seems that PCBs (60.55 ng·g$^{-1}$) had some significant ER effects only on Lake Sirio sediments. A relationship was reported between low ER and DR in Lake Como sediments when DDTs and PCBs were high (55.2 ng·g$^{-1}$ d.w.; 194.5 ng·g$^{-1}$ d.w., respectively) [15]. Ecotoxicological tests on sediments from Lake Como with lower contamination of these substances did not show comparable negative effects.

Regarding what has been said so far, it is useful to underline that contaminants can be present in mixtures in sediments, and two categories of them can be considered only an indication of pollution, since no information on all the present contaminants can be assessed. Even a mitigation effect due to cocktails of unknown contaminants cannot be excluded. This is the reason why the ecotoxicological approach can be an interesting and useful tool to characterize the chemical quality of a matrix, better than single analyses. However, negative effects can be therefore due to contaminants other than those analyzed by water conditions (pH and NH$^{4+}$). Previous work [30] identified pH effects on the survival of eggs and first instar larvae of *Chironomus riparius*, showing the important role of pH when it is lower than 4. observed A lower survival of chironomids was observed in a river where pH decreased to 4.5 [31]. Moreover, studying the effects of ammonium, it was found that 8.0 mg·L$^{-1}$ could be a critical concentration for the survival of chironomids [32]. The negative effects found for the sediments of Sos Canales reservoir and Lake Candia, and in part even in Lake Mergozzo, could be therefore linked to a cocktail of different substances other than DDTs and PCBs.

The effects at the community level as seen with the application of the BQIES are even more complex and not so easily connected with the mere presence of organic pollutants. Chironomids and oligochaetes respond to a wide variety of stressors [33–35]. This is precisely the reason for the use of Biological Quality Elements (BQEs) in the Water Framework Directive (WFD) [1,7,9,36,37]. What was expected from our results was that assessing water quality in the deepest area only and across different depths may provide rather different inferences regarding the ecological quality of a water body as a whole (Table 1). The whole-lake index invariably revealed a scenario of equal, higher, or much higher quality than the index at the bottom of the lake (Table 1). Such discrepancy within the same method revealed that bottom areas, undergoing prolonged periods of anoxia or hypoxia because of eutrophication, frequently showed a highly reduced aquatic life. This is the reason why the BQIES was developed to allow for the separation of the ecological classification of individual sampling stations or individual water layers (epilimnion, hypolimnion, benthic area). In this way, the Environmental Agencies could direct mitigation actions in specific lake areas, notwithstanding that the whole-lake assessment reveals a non-critical situation. One could also speculate that the BQIES, developed to address trophic impacts [10,11], seemed not to be fully reliable to provide a comprehensive idea of the ecological status of a lake if considered alone, and if stressors other than eutrophication are present. Contaminants or other pressures are always possible and prevalent as anthropogenic impact [35,36]; thus, additional indices should be considered to obtain a holistic and potentially reliable view of the ecological status of a lake, mainly when water is used as supply.

## 5. Conclusions

Overall, our results provide empirical evidence to support the idea that risk assessment cannot be easily and reliably performed using one or only a few indices to produce a single number to be used by politicians, managers, and stakeholders to base their decisions on environmental issues. We thus confirm the suggestion of the WFD that biological, chemical, and ecotoxicological approaches should be used together to provide a synergistic and holistic assessment of ecological quality and water security in lakes. Ecotoxicological and chemical indexes (such as EQS) should be improved in order to obtain more useful information when making a risk assessment evaluation, which should include ecological aspects of the lakes' environments.

**Supplementary Materials:** The following are available online at http://www.mdpi.com/2073-4441/12/4/1140/s1, Figure S1. Regions and areas of concern (grey) for the present study, Figure S2. Emergence Rate (%) and Development Ratio of chironomids in the ecotoxicological approach, Table S1. Additional information on sampling localities. Latitude and longitude are in WGS84 system, Table S2. pH, dissolved oxygen (%), and $NH^{4+}$ ($mg{\cdot}L^{-1}$) in water at the start ($T_0$) and end ($T_{end}$) of the ecotoxicological tests with lake sediments belonging to different lakes.

**Author Contributions:** Conceptualization, R.B. and A.B.; methodology, R.B., A.B., D.F. and S.Z.; software, D.F. and S.Z.; validation, D.F. and S.Z.; formal analysis, D.F. and S.Z.; investigation, A.B. and R.B.; resources, A.B. and R.B..; data curation, A.B. and R.B.; writing—original draft preparation, A.B. and R.B.; writing—review and editing, A.B., R.B. and D.F.; visualization, A.B., R.B., D.F. and S.Z.; funding acquisition, A.B. All authors have read and agreed to the published version of the manuscript.

**Funding:** The present research was supported by the EU Life+ Project INHABIT (Local hydro-morphology, habitat and RBMPs: new measures to improve ecological quality in South European rivers and lakes; Contract No: LIFE08 ENV/IT/000413 INHABIT) under the LIFE + Environment Policy and Governance 2008 program.

**Conflicts of Interest:** The authors declare no conflict of interest

## References

1. Birk, S.; Bonne, W.; Borja, A.; Brucet, S.; Courrat, A.; Poikane, S.; Solimini, A.G.; van de Bund, W.; Zampoukas, N.; Hering, D. Three hundred ways to assess Europe's surface waters: An almost complete overview of biological methods to implement the Water Framework Directive. *Ecol. Indic.* **2012**, *18*, 31–41. [CrossRef]

2. Belletti, B.; Rinaldi, M.; Gurnell, A.M.; Buijse, A.D.; Mosselman, E. A review of assessment methods for river hydromorphology. *Environ. Earth Sci.* **2015**, *73*, 2079–2100. [CrossRef]

3. Grizzetti, B.; Pistocchi, A.; Liquete, C.; Udias, A.; Bouraoui, F.; van de Bund, W. Human pressures and ecological status of European rivers. *Sci. Rep.* **2017**, *7*, 3. [CrossRef]

4. Boggero, A.; Zaupa, S.; Fontaneto, D.; Bettinetti, R. The Benthic Quality Index to assess water quality of lakes may be affected by confounding environmental features. *Water* **2020**. under review (same issue).

5. European Union. Directive 2000/60/EC of the European Parliament and of the Council of 23 October 2000 establishing a framework for Community action in the field of water policy. *Off. J. Eur. Union* **2000**, *L327*, 1–73.

6. McFarland, B.; Carse, F.; Sandin, L. Littoral macroinvertebrates as indicators of lake acidification within the UK. *Aquat. Conserv. Mar. Freshw. Ecosyst.* **2010**, *20*, S105–S116. [CrossRef]

7. Poikane, S.; Birk, S.; Böhmer, J.; Carvalho, L.; de Hoyose, C.; Gassner, H.; Hellsten, S.; Kelly, M.; Solheim, A.L.; Olin, M.; et al. A hitchhiker's guide to European lake ecological assessment and intercalibration. *Ecol. Indic.* **2015**, *52*, 533–544. [CrossRef]

8. Reyjol, Y.; Argillier, C.; Bonne, W.; Borja, A.; Buijse, A.D.; Cardoso, A.C.; Daufresne, M.; Kernan, M.; Ferreira, M.T.; Poikane, S.; et al. Assessing the ecological status in the context of the European Water Framework Directive: Where do we go now? *Sci. Total Environ.* **2014**, *497–498*, 332–344. [CrossRef]

9. Solheim, A.L.; Feld, C.K.; Birk, S.; Phillips, G.; Carvalho, L.; Morabito, G.; Mischke, U.; Willby, N.; Søndergaard, M.; Hellsten, S.; et al. Ecological status assessment of European lakes: A comparison of metrics for phytoplankton, macrophytes, benthic invertebrates and fish. *Hydrobiologia* **2013**, *704*, 57–74. [CrossRef]

10. Rossaro, B.; Zaupa, S.; Lencioni, V.; Marziali, L.; Boggero, A. *Indice per la Valutazione della Qualità Ecologica dei Laghi Italiani Basato sulla Comunità Bentonica*; Report CNR-ISE 02.13; Consiglio Nazionale delle Ricerche-Istituto per lo Studio degli Ecosistemi: Verbania Pallanza, Italy, 2013; 13p.

11. Boggero, A.; Zaupa, S.; Cancellario, T.; Lencioni, V.; Marziali, L.; Rossaro, B. *Italian Classification Method for Macroinvertebrates in Lakes. Method Summary*; Report CNR-ISE 03.16; Consiglio Nazionale delle Ricerche-Istituto per lo Studio degli Ecosistemi: Verbania Pallanza, Italy, 2016; 16p.

12. Premazzi, G.; Dalmiglio, A.; Cardoso, A.; Chiaudani, G. Lake management in Italy: The implications of the Water Framework Directive. *Lakes Reserv. Res. Manag.* **2003**, *8*, 41–59. [CrossRef]

13. Fornaroli, R.; Cabrini, R.; Zaupa, S.; Bettinetti, R.; Ciampittiello, M.; Boggero, A. Quantile regression analysis as a predictive tool for lake macroinvertebrate biodiversity. *Ecol. Indic.* **2016**, *61*, 728–738. [CrossRef]

14. Wernersson, A.S.; Carere, M.; Maggi, C.; Tusil, P.; Soldan, P.; James, A.; Sanchez, W.; Dulio, V.; Broeg, K.; Reifferscheid, G.; et al. The European technical report on aquatic effect-based monitoring tools under the Water Framework Directive. *Environ. Sci. Eur.* **2015**, *27*, 1–11. [CrossRef]

15. Bettinetti, R.; Ponti, B.; Quadroni, S. An ecotoxicological approach to assess the Environmental quality of freshwater basins: A possible implementation of the EU Water Framework Directive? *Environments* **2014**, *1*, 92–106.

16. Bettinetti, R.; Giarei, C.; Provini, A. Chemical analysis and sediment toxicity bioassays to assess the contamination of the river Lambro (Northern Italy). *Arch. Environ. Contam. Toxicol.* **2003**, *45*, 72–78. [CrossRef] [PubMed]

17. International Organization for Standardization. *Water Quality. Determination of the Inhibition of the Mobility of Daphnia Magna Straus (Cladocera, Crustacea). Acute Toxicity Test*; ISO 6341-1996; ISO: Geneva, Switzerland, 1996.

18. OECD. *Guideline for the Testing of Chemicals 218. Sediment-Water Chironomid Toxicity Test Using Spiked Sediment*; Organization for Economic Cooperation and Development: Paris, France, 2004.

19. Buraschi, E.; Salerno, F.; Monguzzi, C.; Barbiero, G.; Tartari, G. Characterization of the Italian lake-types and identification of their reference sites using anthropogenic pressure factors. *J. Limnol.* **2005**, *64*, 75–84. [CrossRef]

20. Boggero, A.; Zaupa, S.; Rossaro, B.; Lencioni, V.; Marziali, L.; Buzzi, F.; Fiorenza, A.; Cason, M.; Giacomazzi, F.; Pozzi, S. *Protocollo di Campionamento e Analisi dei Macroinvertebrati Negli Ambienti Lacustri*; MATTM-APAT: Roma, Italy, 2013.

21. Wiederholm, T. (Ed.) Chironomidae of the Holartic region. Keys and Diagnoses. Part I: Larvae. *Entomol. Scand.* **1983**, *19*, 1–457.

22. Timm, T. A Guide to the freshwater Oligochaeta and Polychaeta of Northern and Central Europe. *Lauterbornia* **2009**, *66*, 1–235.

23. Various Authors. *Guide per il Riconoscimento delle Specie Animali delle Acque Interne Italiane, Collana del Progetto Finalizzato 'Promozione della Qualità dell'Ambiente'*; Consiglio Nazionale delle Ricerche: Verona, Italy, 1977–1985; Volumes 29.

24. Bettinetti, R.; Galassi, S.; Guilizzoni, P.; Quadroni, S. Sediment analysis to support the recent glacial origin of DDT pollution in Lake Iseo (Northern Italy). *Chemosphere* **2011**, *85*, 163–169. [CrossRef]

25. Crawley, M.J. *The R Book*, 2nd ed.; John Wiley & Sons Ltd.: West Sussex, UK, 2013.

26. Chapman, D. *Water Quality Assessments: A Guide to Use of Biota, Sediments and Water in Environmental Monitoring*; World Health Organization; University Press: Cambridge, UK, 1996; 651p.

27. Mac Donald, D.D.; Ingersoll, C.G.; Berger, T.A. Development and Evaluation of Consensus-Based Sediment Quality Guidelines for Freshwater Systems. *Arch. Environ. Contam. Toxicol.* **2000**, *39*, 20–31. [CrossRef]

28. Bettinetti, R.; Croce, V.; Galassi, S. Ecological risk assessment for the recent case of DDT pollution in Lake Maggiore (Northern Italy). *Water Air Soil Pollut.* **2005**, *162*, 385–399. [CrossRef]

29. Rousch, J.M.; Simmons, T.W.; Kerans, B.L.; Smith, B.P. Relative acute effects of low pH and high iron on the hatching and survival of the water mite (*Arrenurus manubriator*) and the aquatic insect (*Chironomus riparius*). *Environ. Toxicol. Chem.* **1997**, *16*, 2144–2150. [CrossRef]

30. Orendt, C. Chironomids as bioindicators in acidified streams: A contribution to the acidity tolerance of chironomid species with a classification in sensitivity classes. *Int. Rev. Hydrobiol.* **1999**, *84*, 439–449.

31. Monda, D.P.; Galat, D.L.; Finger, S.E.; Kaiser, M.S. Acute toxicity of ammonia ($NH_3$-N) in sewage effluent to *Chironomus riparius*: II. Using a generalized linear model. *Arch. Environ. Toxicol. Chem.* **1997**, *28*, 385–390. [CrossRef]

32. Hynes, H.B.N. *The Biology of Polluted Waters*; Liverpool University Press: Liverpool, UK, 1960.

33. Brinkhurst, R.O.; Jamieson, B.G.M. *Aquatic Oligochaeta of the World*; Oliver & Boyd: Edinburgh, UK, 1971; 860p.

34. Czechowskia, P.; Stevens, M.I.; Madden, C.; Weinsteing, P. Steps towards a more efficient use of chironomids as bioindicators for freshwater bioassessment: Exploiting eDNA and other genetic tools. *Ecol. Indic.* **2020**, *110*, 105868. [CrossRef]

35. Hering, D.; Carvahlo, L.; Argillier, C. Managing aquatic ecosystems and water resources under multiple stress—An introduction to the MARS project. *Sci. Total Environ.* **2015**, *503–504*, 10–21. [CrossRef]

36. Leavitt, P.R.; Fritz, S.C.; Anderson, N.J. Paleolimnological evidence of the effects on lakes of energy and mass transfer from climate and humans. *Limnol. Oceanogr.* **2009**, *54*, 2330–2348. [CrossRef]

37. Reyjol, Y.; Argillier, C.; Bonne, W.; Borja, A.; Buijse, A.D.; Cardoso, A.C.; Daufresne, M.; Kernan, M.; Ferreira, M.T.; Poikane, S.; et al. Assessing the ecological status in the context of the European Water Framework Directive: Where do we go now? *Sci. Total Environ.* **2014**, *497–498*, 332–344. [CrossRef]

 *water*

*Article*

# The Benthic Quality Index to Assess Water Quality of Lakes May Be Affected by Confounding Environmental Features

Angela Boggero [1],*, Silvia Zaupa [1], Roberta Bettinetti [2], Marzia Ciampittiello [1] and Diego Fontaneto [1]

[1] National Research Council-Water Research Institute (CNR-IRSA), Corso Tonolli 50, 28922 Verbania, Italy; silvia.zaupa@irsa.cnr.it (S.Z.); marzia.ciampittiello@irsa.cnr.it (M.C.); diego.fontaneto@irsa.cnr.it (D.F.)

[2] Department of Human Sciences and Innovation for the Territory (DiSUIT), University of Insubria, Via Valleggio 11, 22100 Como, Italy; roberta.bettinetti@uninsubria.it

*   Correspondence: angela.boggero@irsa.cnr.it

Received: 10 July 2020; Accepted: 6 September 2020; Published: 9 September 2020

**Abstract:** To assess if environmental differences other than water quality may affect the outcome of the Benthic Quality Index, a comparison of the application of four different methods (Benthic Quality Index—BQIES, Lake Habitat Modification Score—LHMS, Lake Habitat Quality Assessment—LHQA and Organisation for Economic Co-operation and Development—OECD) used to classify the lake ecological and hydro-morphological status of 10 Italian lakes was performed. Five lakes were natural and five were reservoirs belonging to both Alpine and Mediterranean Ecoregions. The 10 lakes were sampled using the Water Framework Directive compliant standardized national protocol, which includes sampling soft sediment in the littoral, sublittoral and deep layers along transects with a grab of 225 cm$^2$ during spring and autumn. The application of Generalised Linear Mixed Effect Models both at the lake level and at the single station of each lake highlighted that, at the lake level, no significant correlations existed between any couple of hydro-morphological, ecological and trophic status assessments, with each metric representing a different facet of human impact on the environment. At the single site level, we found significant effects of depth on the metrics of biodiversity. The best approximation of single-site macroinvertebrates diversity among the metrics of overall lake quality was with the LHMS, but not with the BQIES. Our hypotheses that lake macroinvertebrates assemblages depend also on other potential confounding variables of habitat degradation and intrinsic differences between lakes were confirmed, with depth playing a major role. Therefore, the assessment of lakes with different depths may produce different whole-lake BQIES values, only because of the effect of depth gradient and not because of differences in lake quality.

**Keywords:** grain-size; sediment; chemical analysis; macroinvertebrates; ecological status; Water Framework Directive; multimetric indices

---

## 1. Introduction

Before the launch of the Water Framework Directive (WFD, Directive 2000/60/EC), due to the extensive use of waters for indoor and outdoor purposes (e.g., hydropower generation, domestic, agricultural, industrial and recreation scopes), several of the aquatic ecosystems in Europe were heavily degraded, and many of them completely lost, sometimes even in an irreversible way [1,2]. Thus, the WFD is an important component in supporting the water sector in Europe, emphasizing the role of aquatic ecology in management decisions to protect an exhaustible resource [3]. Water resources management is based on a comprehensive understanding of ecosystem functions and interactions, so a multi-parametric approach was needed to sustain water policy at the European level, considering future conservation and restoration actions [4].

In order to improve the quality of surface water bodies (lakes and rivers), specific studies focused on the implementation of monitoring and assessment methods across Europe. The most prominent are:

- Screening methods for Water Data Information in support of the implementation of the Water Framework Directive (SWIFT-WFD: www.swift-wfd.com) [5];
- Development and Testing of an Integrated Assessment System for the Ecological Quality of Streams and Rivers throughout Europe using Benthic Macroinvertebrates (AQUEM: www.aqem.de) [6];
- Standardisation of River Classifications: Framework method for calibrating different biological survey results against ecological quality classifications to be developed for the Water Framework Directive (STAR: www.eu-star.at/frameset.htm) [7];
- Relationships between the ecological and chemical status of surface waters (REBECCA: www.ymparisto.fi/eng/research/euproj/rebecca/homepage.html) [8];
- Water Bodies in Europe: Integrative systems to assess ecological status and recovery (WISER: www.wiser.eu) [9];
- Local hydro-morphology, habitat and RBMPs: new measures to improve ecological quality in South European rivers and lakes (INHABIT: www.life-inhabit.it);
- Managing aquatic ecosystems and water resources under multiple stresses (MARS: www.mars-project.eu) [10,11].

The evaluation of the ecological status of a lake requires an integrated approach that takes into account the effects on biota of different pressures encountered in lakes (eutrophication, acidification, general degradation, morphological alteration, etc.). As a general rule, the composition and the characteristics of the biotic communities of European waters, and the abiotic conditions influencing them, have become a primary focus to be analyzed in order to assess the quality of lakes and rivers as a whole. Since 2000, several biological assessment metrics, biotic indices or predictive models covering taxonomic, functional and trait-based approaches considering different pressures were developed by various countries in Europe [12–21], with the aim of improving management and conservation actions throughout Europe, and of harmonizing the classification of ecological status. Italy, after recognizing of the importance of eutrophication as a pressure impact on the national territory (nearly 41% of the Italian lakes were eutrophic [22]), decided to assess eutrophication impacts in lakes. This goal was reached using specific indices, ecological quality ratios and chemical and hydro-morphological status assessment to define reference conditions. Different metrics for phytoplankton, diatoms, macrophytes, macroinvertebrates and fish were developed within the remits of the WFD (for details on the Italian metrics adopted for each Biological Quality Element (BQE), see www.ise.cnr.it/wfd-en). Among such metrics, the Benthic Quality Index (BQIES) [23,24] considers the composition of the macroinvertebrates assemblages in order to assess the eutrophication levels of lakes. The index is based on a species-level approach for all benthic macroinvertebrates, mainly for chironomids and oligochaetes, co-dominating lake benthic communities. Then, in a second step, using quantile regression analysis, a rapid bio-assessment methodology of quality conditions has been set up to be submitted to the authorities responsible for water monitoring and to water managers [25]. The application of the rapid bio-assessment methodology has the objective of optimizing the sampling procedures of the national standardized protocol for monitoring lakes [26]. The aim is to support the environmental agencies responsible for the assessment of ecological quality in identifying entire lakes or parts of them that are altered, turning their attention to them and starting remediation actions. In 2018, the BQIES was finally accepted at the European level (UE Decision 2018/229) and became fully operational at the end of the same year.

The BQIES, like other indices, should reflect the effect of pollution on water quality, but also the effects of the physical, chemical, biological and biogeographic characteristics of each water body [27]. The explicit assessment of most of these environmental features is not currently compulsory in the application of the BQIES [24], but may be used as a further support in the definition of high ecological status. We here want to demonstrate that the nature of the sediment, the lake depth and the water chemistry, and not only the trophic status, are factors affecting the outcome of the BQIES [28–31]. This paper tests the hypothesis that macroinvertebrates, sampled using a standardized methodology over a short period of time (spring to autumn of the same year) in both natural lakes and reservoirs in Italy, respond to a gradient of trophic state in which agriculture and animal husbandry are the predominant stressors, according to the current use of the BQIES, but also to environmental variables that could represent confounding factors for the strict application of the BQIES. To assess if environmental differences other than water quality may affect the outcome of the BQIES, we performed a comparison of the application of four different methods (BQIES, Lake Habitat Modification Score—LHMS, Lake Habitat Quality Assessment—LHQA, Organisation for Economic Co-operation and Development—OECD) used to classify lake ecological and hydro-morphological status. Then, we analyzed how the variability in species compositions of lake macroinvertebrates is related to geographic and local scale environmental factors, including sediment texture, sediment organic and inorganic matter content, hydro-chemical conditions and depth, in comparison with other indices of environmental degradation. The outcome of our analyses could be used to improve the application of the BQIES and to make it more compliant to the aims of the WFD, clarifying the importance of local environmental parameters not only as a support tool, but as a key means of characterizing the sediment on which life within them depends.

## 2. Materials and Methods

### 2.1. Study Area

The WFD states that all lakes with a surface area >0.5 km$^2$ should be monitored [22], but administrative regions, or parks or water managers, can decide to include smaller lakes deserving particular protection and safeguarding because of their strategic importance as a drinking water supply, their particular environmental value, or the peculiar character of the fauna and flora inhabiting them [32].

The lakes included in the current study belong to different types according to the EC Water Framework Directive classification system [33], and to two separate Ecoregions, together covering the meteorological conditions typical of the whole country. The lakes are classified into Alpine (AL) and Mediterranean lakes (ME) on the basis of the Ecoregion agreement [33]. The choice of this set of lakes and reservoirs was based on previous information about their trophic state, allowing us to cover a gradient of trophic state (from ultra-oligotrophy to hypertrophy) within which to test our hypotheses.

Six lakes are located in continental Italy (north-western side, Piedmont) and the remaining four in insular Italy (Sardinia) (Figure 1). They thus are subject to different meteo-climatic conditions. Five lakes are natural and five are reservoirs, of which four are placed in rural areas and one in natural settings (Table 1).

**Figure 1.** Distribution of considered administrative regions and sampled lakes in Italy.

**Table 1.** Administrative regions, lake names and types, ecoregions (AL = Alpine, ME = Mediterranean) and class types of lakes, and their respective main geographic and morphological information. Latitude and longitude, in WGS84 system, are expressed as DMS (Degrees, Minutes, Seconds); Altitude: m a.s.l.; Lake area: km$^2$; Mean and Maximum depth: m.

| Region | Lake Name | Lake Type | Class Type | Latitude N | Longitude E | Altitude | Lake Area | Mean Depth | Maximum Depth |
|---|---|---|---|---|---|---|---|---|---|
| Piedmont | Avigliana piccolo | Natural | AL-5 | 45°03′13″ | 07°23′30″ | 356 | 0.58 | 7.70 | 12.00 |
| Piedmont | Candia | Natural | AL-5 | 45°19′25″ | 07°54′43″ | 227 | 1.69 | 5.90 | 8.00 |
| Piedmont | Mergozzo | Natural | AL-6 | 45°57′23″ | 08°27′47″ | 194 | 1.82 | 45.60 | 73.00 |
| Piedmont | Morasco | Reservoir | AL-9 | 46°25′33″ | 08°23′48″ | 1814 | 0.57 | 31.00 | 50.00 |
| Piedmont | Sirio | Natural | AL-6 | 45°29′06″ | 07°53′05″ | 271 | 0.32 | 18.10 | 44.00 |
| Piedmont | Viverone | Natural | AL-6 | 45°24′05″ | 08°03′05″ | 230 | 5.78 | 22.50 | 50.00 |
| Sardinia | Bidighinzu | Reservoir | ME-2 | 40°33′24″ | 08°39′44″ | 330 | 1.50 | 8.40 | 30.00 |
| Sardinia | Liscia | Reservoir | ME-4 | 40°59′39″ | 09°14′37″ | 177 | 5.57 | 18.80 | 63.50 |
| Sardinia | Posada | Reservoir | ME-3 | 40°38′19″ | 09°36′28″ | 43 | 3.00 | 9.30 | 29.50 |
| Sardinia | Sos Canales | Reservoir | ME-5 | 40°33′17″ | 09°18′55″ | 709 | 0.30 | 19.70 | 47.50 |

## 2.2. Lake Classification

Water bodies (natural lakes and reservoirs) were sampled for biota and physical and chemical analyses in spring and autumn along transects connecting the littoral, sub-littoral and deep areas of each lake, according to the national protocol for macroinvertebrates sampling [26].

Four different methods were used to classify lake ecological and hydro-morphological status: OECD, LHMS, LHQA and BQIES. The methodology proposed by the Organisation for Economic Co-operation and Development (OECD [34]) was used to classify lakes on the basis of their trophic conditions. This metric is based on the hypolimnic oxygen concentrations in deep layers, on the mean values of total phosphorus (TP) at mixing, on mean values of chlorophyll *a*, and on mean annual values of transparency. TP, oxygen and chlorophyll *a* analyses were performed following Tartari and Mosello [35], and transparency was measured with a Secchi disc.

The summer application of the Lake Habitat Survey (LHS) was used to characterize the hydro-morphological conditions of each lake [36]. The LHS, based on a combination of habitat plot (Hab-Plots) observations, generates two main summary metrics: LHMS (Lake Habitat Modification Score), related to the degree of site modification, and LHQA (Lake Habitat Quality Assessment), to measure the diversity and naturalness of the lakes. Both metrics in the LHS assessment method, which surveys the terrestrial/aquatic ecotone, include quantitative descriptions of vegetation canopy, macrophytes composition and distribution, main littoral substrate and the presence of human impacts on the shores and riparian zone [36]. High values of LHMS indicate high human modification and thus low ecological status, whereas high values of LHQA indicate high naturalness and thus high ecological status or high habitat quality.

The Benthic Quality Index (BQIES) was applied to the lakes as well. It is based on indicator weights attributed to each species, assuming that a species that is known to live preferably in high diversity sites is an indicator of a healthy environment, whereas a species that is known to be abundant in low diversity sites is an indicator of altered environments. Thus, high values of the index indicate high biodiversity and high ecological status [24].

## 2.3. Sampling Methodology

Biota and soft sediment were sampled with a grab (area = 225 cm$^2$); biota was sieved in the field through a net (mesh 250 µm), and stored with an aqueous solution (5%) of buffered formaldehyde. Water samples were collected by a Niskin bottle equipped with an overturning thermometer to obtain water parameters from the different sampling depths. All samples were brought to the laboratory for subsequent analyses.

Water features were measured for each sampling point in each lake according to Fornaroli and co-authors [25], and include both physical and chemical metrics: temperature, oxygen concentration, pH, conductivity, alkalinity, total phosphorus (TP) and total nitrogen (TN) (Table S1).

Sediment features were analyzed regarding chemistry and texture. For sediment chemistry, the water content, organic and inorganic matter and percentage of carbonates were measured via Loss On Ignition analysis (LOI 500 °C—[37]). For sediment texture, grain size was analyzed via Wentworth scaling (U.S. Standard) [38], allowing the separation of the sediment into many different fractions below 2 mm [39,40], and their classification according to their constituent parts (clay, silt and fine sand, expressed as percentages).

In the lab, sediment samples were sorted under a stereomicroscope to identify macroinvertebrates. The animals were separated into main groups, identified to species level when the presence of juveniles did not preclude it, and counted. The identification manuals used are those in use at the international and national levels ([41] for Chironomids; [42] for Oligochaetes, and [43] for the remaining taxonomic groups). Richness (number of taxa, considering mostly species, but also genera and families for minor taxonomic groups) and diversity (Shannon diversity index—SDI), representing the community structure and complexity, were calculated for each single-site sampling point.

The biological data for all taxa identified among the macroinvertebrates were used to apply the BQIES to each sampling point within a lake (BQIES$_{single-site}$—[24]). For each single-site sampling point we obtained biological and abiotic measurements.

In addition to the analyses performed at each sampling point, we also followed the current regulations, averaging values of all sampling points through space and time to obtain a mean annual BQIES$_{whole-lake}$ value for each lake.

## 2.4. Statistical Analyses

The first series of analyses involved a comparison between classification systems at the lake level; we compared BQIES$_{whole-lake}$ (whole-lake assessment of the BQIES), OECD, LHMS and LHQA with Pearson multiple correlations, using R v 4.0.0 [44], package psych v1.9.2 [45].

The second series of analyses assessed the potential influence of depth on the other environmental metrics, including sediment texture, sediment chemistry, and the water physical and chemical parameters of each single-site sample. Our hypothesis was that the nature of sediment and water features would affect macroinvertebrates, and if the sediment and water features change with depth, space and time, the cascading effects on the macroinvertebrates assemblages could affect the BQIES$_{whole-lake}$ assessment. To verify our hypothesis, we used Generalized Linear Mixed Effect Models (GLMMs [46]), analyzing the effect of depth on the variability of the environmental metrics, accounting for the pseudoreplication within each lake of the random effects, and for the effect of seasonality as an additional explicit factor in the model. GLMMs were performed with the R package nlme v3.1-147 [47].

After exploring the lake classification and abiotic features of the sampling sites, the main question we wanted to address was the potential confounding effect of sediment and water features in driving the differences in species assemblages of macroinvertebrates in comparison to the representativeness of macroinvertebrates as BQE for the lake quality assessment. We addressed this question by using GLMMs on macroinvertebrates richness, macroinvertebrates SDI and BQIES$_{single-site}$ applied to each site as a function of depth, sediment chemistry, sediment texture, water features and lake ecological status classifications (according to OECD, LHMS, LHQA and BQIES$_{whole-lake}$), accounting for the effect of seasonality and including the effect of the pseudoreplication of sites nested within each lake as a random effect. For the statistical models we used one single explanatory variable summarizing sediment chemistry, sediment texture and water features. To obtain such summary metrics, we performed principal component analyses, one for each group of variables, and kept the first axis as a summary of the metrics of the group.

Before any analysis, dependent variables expressed as percentages were arcsine square root transformed, and dependent variables expressed as count data were log transformed to obtain a Gaussian distribution of residuals [48]. For models including a combination of continuous variables and categorical variables with more than two levels, summary outputs were obtained as type II analysis of deviance tables with the R package car v3.0.7 [49]. Partial $r^2$ for GLMMs were obtained with the R package r2glmm v0.1.2 [50].

## 3. Results

### 3.1. Lake Classification

The range of averaged BQIES$_{whole-lake}$ values from each lake varied between 0.52 (Lake Mergozzo) and 0.22 (Lake Sirio); most of the lakes' estimates were lower than 0.4, a threshold between good (higher than 0.4) and moderate status (Table 2).

**Table 2.** Summary metrics and classification for each analyzed lake, including annual mean hypolimnic oxygen saturation during stratification (O2, %), annual mean total phosphorus at mixing (TP, mg m$^{-3}$), annual mean chlorophyll *a* (Chl *a*, mg m$^{-3}$), and annual mean transparency (m), used to obtain the OECD classification; the other classification scores include BQIES$_{whole-lake}$, LHMS and LHQA.

| Lakes | O$_2$ | TP | Chl *a* | Transparency | OECD | BQIES$_{whole-lake}$ | LHMS | LHQA |
|---|---|---|---|---|---|---|---|---|
| Avigliana piccolo | 3 | 18.56 | 2.17 | 4.17 | oligo-mesotrophic | 0.255 | 26 | 56 |
| Bidighinzu | 3 | 259.16 | 14.67 | 0.89 | hypertrophic | 0.391 | 20 | 47 |
| Candia | 19 | 27.70 | 11.02 | 2.74 | eutrophic | 0.421 | 26 | 56 |
| Liscia | 15 | 41.97 | 6.73 | 3.04 | meso-eutrophic | 0.268 | 18 | 61 |
| Mergozzo | 75 | 4.60 | 1.98 | 7.50 | oligotrophic | 0.518 | 14 | 56 |
| Morasco | 78 | 2.20 | 0.36 | 7.33 | ultra-oligotrophic | 0.464 | 26 | 30 |
| Posada | 32 | 37.44 | 7.35 | 1.97 | eutrophic | 0.370 | 24 | 48 |
| Sirio | 3 | 41.07 | 3.70 | 5.03 | meso-eutrophic | 0.220 | 32 | 53 |
| Sos Canales | 6 | 31.67 | 6.66 | 2.80 | eutrophic | 0.331 | 26 | 52 |
| Viverone | 59 | 80.13 | 3.42 | 5.62 | meso-eutrophic | 0.255 | 32 | 62 |

When classifying Italian lakes on the basis of OECD trophic conditions (Table 2), Lake Mergozzo was the only one with a high annual mean oxygen content even in the deepest layers, low mean concentrations of TP and chlorophyll *a*, and high annual mean transparency values.

The natural lakes of the present study, through the application of LHS (Table 2), were characterized via high habitat quality (mean LHQA ± SD = 56.6 ± 3.3). In two cases (lakes Viverone and Sirio), the lake modification measures (LHMS) were the highest because of the pronounced human littoral alterations. Once again, Lake Mergozzo showed very good hydro-morphological conditions with the lowest LHMS (14) and a quite high LHQA (56) (Table S1), confirming the high habitat quality and conservation value of its habitats. The reservoirs showed LHMS values never exceeding 20 (Table 2), with the dam representing the only adverse environmental impact.

Each system of lake classification provided a different facet of the assessment of anthropogenic impacts on the habitat. The pairwise correlation values between classification metrics were very low, always below 0.6 as an absolute value, and never significant (Figure 2).

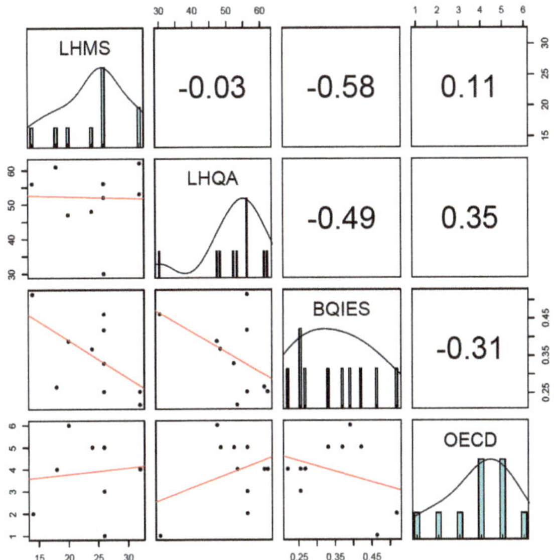

**Figure 2.** Pairwise comparisons between classification systems: LHMS for habitat modification, LHQA for naturalness, BQIES for macroinvertebrates and OECD for trophic status, ranked in order from 1 as ultra-oligotrophic to 6 as hypertrophic. The diagonal reports the histograms of the distribution of each metric; the values above the diagonal represent Pearson's r correlation values; the scatterplots below the diagonal show the correlation between pairs of variables with the (non-significant) trend line.

## 3.2. Single-Site Sediment and Water Descriptors

The sediment chemical components (Table S1) revealed that Morasco reservoir had the lowest average water content (38%) and organic matter percentages (3%); the highest organic matter contents were found in lakes Candia (48%), Sirio (35%) and Viverone (23%), whereas values were <20% in the other lakes (Figure 3). Carbonates were usually present in smaller amounts than the other analyzed components, with values ranging on average from 1% (Sos Canales reservoir) to 20% (Lake Avigliana piccolo).

Sediment texture analyses (Table S1) revealed fine sand as the main component (Figure 3), while silt and clay represented only minor fractions. Morasco reservoir and Lake Candia represented an exception since they showed higher values for both silt and clay. In detail, the lowest fine sand content was found in the Morasco reservoir (60%), and the highest in Lake Sirio (92%). Silt ranged from 6% (Lake Sirio) to 21% (Lake Candia), whereas clay varied from 1% (Liscia reservoir) to 20% (Morasco reservoir).

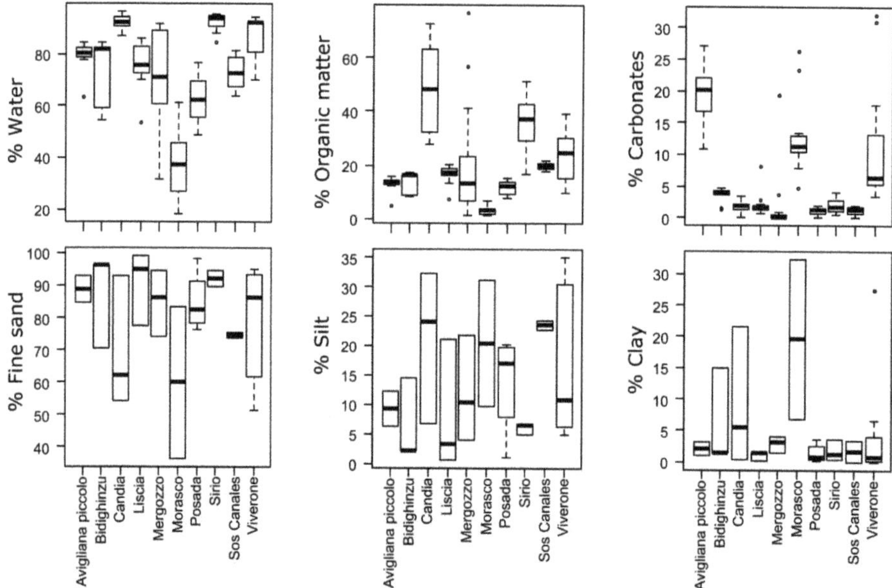

**Figure 3.** Distribution of sediment chemical features (up: water, organic matter and carbonates contents) and soft sediment texture (down: fine sand, silt, clay) for each lake. All variables are reported as percentages.

The water features (Table S1) showed a wide range of conductivity, from 56 (Lake Mergozzo) to >400 μS cm$^{-1}$ (Lake Avigliana piccolo and Bidighinzu reservoir), and alkalinity (from 14 mg L$^{-1}$ in Lake Mergozzo to 398 mg L$^{-1}$ in Lake Avigliana piccolo). The pH values ranged between 6.5 and 9.1, with the lowest values (<7) in the deep areas of lakes Mergozzo and Sirio and the Sos Canales reservoir, and the highest values (>8) in the littorals of the lakes in north-western Italy. Oxygen saturation was highly variable (from 1.1 to 128%) and strictly dependent on depth and season; the highest concentrations were found along the shores of Lake Sirio, and very low values were found in the deepest layers of Lake Viverone, as well as the Bidighinzu, Sos Canales and Liscia reservoirs, which were close to anoxia (<5%) during prolonged period of water stratification. The nutrient conditions (both TP and TN) showed again a wide range of values: the former varied from 4 μg L$^{-1}$ (Lake Mergozzo) to 1081 μg L$^{-1}$ (Lake Bidghinzu), the latter from 2.1 mg L$^{-1}$ (Lake Avigliana piccolo) to >3 mg L$^{-1}$ (Lake Bidighinzu).

Some of the sediment and water features of the samples revealed a significant relationship with depth (Table 3). Among the metrics describing sediment chemistry, only water content was significantly and positively related to depth; among the metrics describing sediment texture, all the percentages of fine sand, silt and clay were significantly and positively related to depth; among water features, temperature, oxygen, pH, TP and TN were significantly affected by depth (Table 3).

**Table 3.** Output of the Generalized Linear Mixed Effect Models (GLMM) assessing the effect of depth on sediment chemistry, sediment texture and water features, including seasonality as a covariate and sites nested within each lake as random effects. Significant effects of depth are marked in bold.

| Environmental metrics | Parameters | Predictor | t-Value | p-Value |
|---|---|---|---|---|
| sediment chemistry | water content | (intercept) | 12.6 | <0.001 |
| | | depth | 2.9 | **0.007** |
| | | season | 0.6 | 0.547 |
| | organic | (intercept) | 5.2 | <0.001 |
| | | depth | −1.2 | 0.238 |
| | | season | −0.1 | 0.965 |
| | inorganic | (intercept) | 17.0 | <0.001 |
| | | depth | 1.6 | 0.114 |
| | | season | 1.1 | 0.299 |
| | carbonates | (intercept) | 3.5 | 0.002 |
| | | depth | −1.0 | 0.329 |
| | | season | −2.3 | 0.031 |
| sediment texture | fine sand | (intercept) | 22.0 | <0.001 |
| | | depth | 3.1 | **0.005** |
| | | season | −6.6 | <0.001 |
| | silt | (intercept) | 12.3 | <0.001 |
| | | depth | −2.7 | **0.011** |
| | | season | −5.5 | 0.001 |
| | clay | (intercept) | 3.1 | <0.001 |
| | | depth | −1.6 | **0.019** |
| | | season | −0.1 | 0.598 |
| water features | temperature | (intercept) | 12.5 | <0.001 |
| | | depth | −5.3 | **<0.001** |
| | | season | 3.4 | 0.002 |
| | oxygen | (intercept) | 9.5 | <0.001 |
| | | depth | −3.6 | **0.002** |
| | | season | −1.5 | 0.140 |
| | pH | (intercept) | 49.7 | <0.001 |
| | | depth | −3.1 | **0.005** |
| | | season | −0.1 | 0.940 |
| | conductivity | (intercept) | 5.2 | <0.001 |
| | | depth | 1.5 | 0.142 |
| | | season | 1.5 | 0.146 |
| | alkalinity | (intercept) | 3.3 | 0.002 |
| | | depth | 0.9 | 0.358 |
| | | season | 0.3 | 0.767 |
| | TP | (intercept) | 0.9 | 0.334 |
| | | depth | 2.2 | **0.035** |
| | | season | 0.5 | 0.614 |
| | TN | (intercept) | 4.5 | <0.001 |
| | | depth | 2.4 | **0.023** |
| | | season | −0.9 | 0.361 |

*3.3. Whole-Lake Macroinvertebrates Assemblages*

Macroinvertebrates were represented by 12,799 individuals, belonging to 142 taxa in seven classes (Arachnida, Bivalvia, Clitellata, Gastropoda, Insecta, Platyhelminthes and Malacostraca) and 36 families. Oligochaetes and Chironomids were the most abundant groups, whose overall relative abundances constituted from 46% (Lake Candia) to 100% (Morasco reservoir) of the macroinvertebrates assemblage (Figure 4).

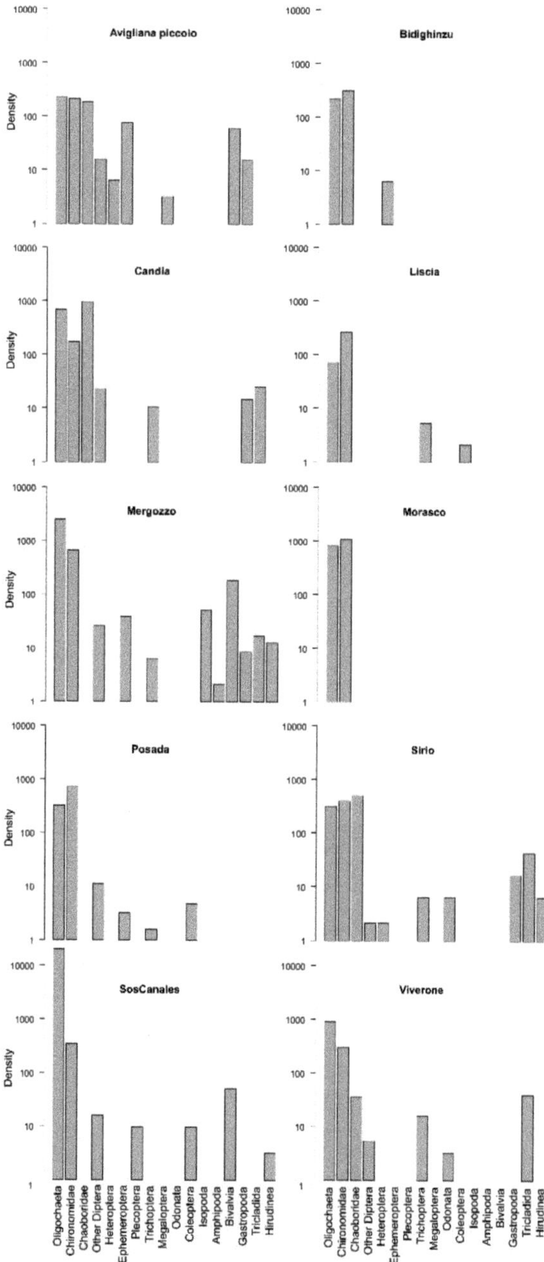

**Figure 4.** Logarithmic (ln) distribution of densities (ind m$^{-2}$) of the main macroinvertebrates taxonomic groups in the different lakes.

Chironomids tended to prevail over Oligochaetes in most lakes, while oligochaetes prevailed in lakes Mergozzo and Viverone, and the Sos Canales reservoir. The latter showed the highest oligochaetes absolute abundances (around 20,000 ind m$^{-2}$), whereas lakes Mergozzo and Viverone had absolute abundances of 2400 ind m$^{-2}$ and 900 ind m$^{-2}$, respectively. Lakes Sirio and Candia presented a large number of chaoborids (550 and 900 ind m$^{-2}$, respectively), while Lake Avigliana piccolo showed comparable densities of oligochaetes, chironomids and chaoborids (225, 216 and 184 ind m$^{-2}$, respectively), and relatively high densities for mayflies, bivalves and gastropods (76, 60 and 16 ind m$^{-2}$, respectively).

### 3.4. Environmental Effects on Macroinvertebrates

Each group of environmental features, namely sediment chemistry, sediment texture and water features, for each site was summarized in one single axis of a principal component analysis, which produced an explained variance 89.8% for sediment chemistry, 77.6% for sediment texture, and 68.3% for water features. These axes were included in the following models as proxies for the three environmental features.

The statistical models to assess the role of potential confounding factors on metrics of diversity (richness of macroinvertebrates, SDI and BQIES$_{single\text{-}site}$) revealed that depth had a significant effect, and that, among the metrics of lake quality, LHMS was always significant (Table 4).

**Table 4.** Type II analysis of deviance tables as output of the GLMM assessing the effect of sediment chemistry, sediment texture, water features and depth, including seasonality as a covariate, in addition to the metrics of overall lake quality (OECD, LHMS, LHQA and BQIES$_{whole\text{-}lake}$) and sites nested within each lake as random effects, on three metrics of site biodiversity: richness, SDI (diversity) and BQIES$_{single\text{-}site}$. Significant effects are marked in bold. The model $R^2$ is reported for each model between parentheses after the name of the response variable.

| Response | Predictor | $\chi^2$ | $p$-Value | $R^2$ |
|---|---|---|---|---|
| richness ($R^2 = 0.81$) | PC axis sediment chemistry | 0.0 | 0.9587 | 0.00 |
| | PC axis sediment texture | 0.7 | 0.3903 | 0.02 |
| | PC axis water features | 0.4 | 0.5486 | 0.01 |
| | depth | 27.5 | **<0.0001** | 0.60 |
| | OECD | 0.1 | 0.7851 | 0.00 |
| | LHMS | 9.5 | **0.0021** | 0.28 |
| | LHQA | 3.8 | 0.0512 | 0.13 |
| | BQIES$_{whole\text{-}lake}$ | 0.1 | 0.7200 | 0.01 |
| | season | 5.1 | 0.0768 | 0.04 |
| SDI ($R^2 = 0.70$) | PC axis sediment chemistry | 2.0 | 0.1592 | 0.09 |
| | PC axis sediment texture | 0.3 | 0.6050 | 0.01 |
| | PC axis water features | 0.2 | 0.6789 | 0.01 |
| | depth | 24.6 | **<0.0001** | 0.55 |
| | OECD | 0.6 | 0.4214 | 0.04 |
| | LHMS | 5.3 | **0.0209** | 0.21 |
| | LHQA | 0.0 | 0.8764 | 0.00 |
| | BQIES$_{whole\text{-}lake}$ | 0.1 | 0.8197 | 0.00 |
| | season | 3.3 | 0.1964 | 0.04 |
| BQIES$_{single\text{-}site}$ ($R^2 = 0.78$) | PC axis sediment chemistry | 5.5 | **0.0191** | 0.26 |
| | PC axis sediment texture | 0.0 | 0.8685 | 0.00 |
| | PC axis water features | 0.0 | 0.9900 | 0.00 |
| | depth | 20.7 | **<0.0001** | 0.57 |
| | OECD | 0.5 | 0.4872 | 0.03 |
| | LHMS | 6.2 | **0.0131** | 0.22 |
| | LHQA | 0.1 | 0.7734 | 0.00 |
| | BQIES$_{whole\text{-}lake}$ | 0.0 | 0.8681 | 0.00 |
| | season | 8.3 | **0.0159** | 0.05 |

Depth explained between 55% and 60% of the variance for each model (Table 4), and LHMS between 21% and 28%. None of the other predictors were relevant to explaining richness and SDI, whereas for BQIES$_{single-site}$, sediment chemistry (water content) and seasonality were also significant. Surprisingly, the BQIES$_{single-site}$ scores were not related to the overall BQIES$_{whole-lake}$ scores, nor to the trophic status assessed according to OECD standards (Table 4).

## 4. Discussion

The first unforeseen and positive result of our study at the lake level was that no significant correlations existed between any couple of hydro-morphological, ecological and trophic status assessments (LHQA, LHMS, BQIES$_{whole-lake}$ and OECD) (Figure 2). One explanation could be that each metric assesses a different facet of the environmental impacts of anthropogenic activities, using as they do the descriptive morphologic and hydrologic characteristics of the littorals, the biodiversity of macroinvertebrates or nutrients, and that the different facets of human impacts on different environmental features are not strictly correlated but actually complementary [51,52]. Such differences have two consequences: on the one hand, the different metrics may all be necessary for a reliable implementation of the WFD; on the other hand, ecological assessments cannot be provided with a single number that evaluates lake quality. The Water Framework Directive considers healthy ecosystems to be the basis of sustainable water resources, whereby the various components are interconnected and provide ecosystem services with positive cascading effects on lake resilience to counteract short-term impacts. This means that each assessment, representing different pressures and impacts, did not provide any redundancy in the ecological assessment of lakes, but rather that each of them constitutes a description of the environmental complementarity of the ecological status of each lake. This would have potential implications on mitigation actions to be taken: different metrics of human impacts could already suggest which actions should be considered to minimize impacts on morphological, hydrological, chemical or biodiversity features, understanding their effectiveness in avoiding threats to the provision of ecosystem services on which humanity depends [53].

The other results of our analysis were in line with our expectations: sediment and water features indeed may change with depth, even within the same lake, thus differentially affecting macroinvertebrates assemblages on top of the ecological quality of the lake, and potentially providing biased assessments of lake quality through indices that do not consider such confounding factors. The effect of depth on water and sediment features was not due simply to the fact that the dataset included both deep and shallow water bodies, deeper than 15 m or not; by repeating the same analyses removing the two shallowest lakes (Avigliana piccolo and Candia), the effect of depth was still significant for several water features, namely temperature, oxygen, pH and TN, but not conductivity, alkalinity or TP (Table S2), similar to what happened with the analyses on the overall dataset (Table 3). Regarding sediment, by removing the two shallower lakes, no changes could be seen in the effect of depth on sediment texture and sediment chemistry (Table S2).

The benthic macroinvertebrates species composition in lakes is known to vary with space and time depending on natural factors [54]. As a general rule, environmental variables and biotic interactions influence the macroinvertebrates assemblages, the presence and ratio of sensitive to tolerant species, and the ecological functioning of lakes and their productivity [55–59]. In the present study, the analyses applied to sediment chemistry, sediment texture and water features allowed us to highlight the effect of them, and especially of depth in explaining variations within each lake. This is also in agreement with different metrics of biodiversity, including richness and diversity, confirming the importance of depth in determining the distribution of benthic macroinvertebrates assemblages, and consequently of the BQIES$_{single-site}$, on which basis the BQIES$_{whole-lake}$ is then calculated. Even when removing the two shallowest lakes (Avigliana piccolo and Candia) from the analyses, the effect of depth was still highly significant, and none of the other variables became significant (Table S3). Thus, we can exclude the interpretation that the effect of depth in our study was due to the inclusion of both deep and shallower water bodies, and the effect holds true even in the analyses including only the subset of deeper water bodies.

The lakes and reservoirs we analyzed presented a fairly similar sediment texture because we focused our monitoring program on soft sediment, and such lack of great variability may explain why no effect of sediment texture was found on the species richness, SDI or BQIES$_{single-site}$ of benthic macroinvertebrates. Watershed characteristics influence ecological activities and equilibria by controlling the chemistry of soils [60], plants [61], waters [62] and microbial community composition [63]. Thus, similar sediments may host communities of benthic macroinvertebrates with similar diversity metrics even if they come from different ecoregions, with different origins of the sediment (e.g., sediment of glacial origin is only found in the Alps). Thus, because of such low variability in sediment texture in the analyzed water bodies, we cannot state whether the inclusion of sediment characterization in terms of the percentage of sand, silt and clay, actually not compulsory within the standardized sampling protocol adopted at national level [26], could be useful when monitoring macroinvertebrates in order to facilitate the interpretation of the results. In our case, the differences in depth overruled any smaller difference in sediment texture, changes in which were also directly correlated to depth (Table 3).

An unexpected result of our analysis was that the BQIES$_{single-site}$, as well as species richness and SDI, were not related to the BQIES$_{whole-lake}$, but were more strongly explained by LHMS, as an index of habitat modification. Thus, the inference we can provide is that macroinvertebrates biodiversity within the lake was affected by visible shore modifications, as evaluated by the LHMS, affecting even the deepest parts of the lakes. Such effect was maintained even when removing the shallowest lakes (Table S3), suggesting that LHMS, contrary to what was found by McGoff and co-authors [64], could indeed provide a reliable metric for the assessment of overall lake quality. To provide some additional speculation in support of the reliability of LHMS as an overall whole-lake assessment, or as a single-site metric, we repeated the analyses using the values of LHMS and LHQA measured for each single-site instead of their whole-lake counterparts, including only the littoral or sublittoral sites in each lake corresponding to the sites where LHMS and LHQA were actually measured. The results revealed that the single-site LHMS and LHQA were never significant, either for richness or for SDI or BQIES$_{single-site}$, and could never explain more than 5% of the variance for each model, with depth remaining the most significant predictor (Table S4). Such additional analyses confirm the validity and reliability of LHMS as a whole-lake assessment of human lake-shore modifications, and their effects on the aquatic macroinvertebrates, regardless of the singe-site measurements for the same index.

## 5. Conclusions

Our results confirm the hypothesis that lake macroinvertebrates assemblages are correlated with confounding factors in each lake, such as sediment chemistry and texture and water features, but that most of the variability can be explained by depth, at least in the set of analyzed water bodies. The main inference that can be suggested from our study is that the BQIES based on the standardized monitoring protocol remains a useful tool for the ecological assessment of the quality of deep Italian lakes, with a mean depth higher than 15 m, as previously suggested [24], but also that, actually, no difference in its results could be highlighted between deeper and shallower lakes. The proposed BQIES cannot be considered definitive, as new species with different auto-ecological requirements could potentially be collected in lakes not sampled yet in the central and southern part of Italy. The indicator weights of species are actually based on a historical dataset of geographic distribution, and need to be updated as monitoring proceeds, slightly influencing the outcome of the BQIES assessment.

Another general inference is the support for the reliability of LHMS as an index of overall lake quality, reflected as a significant predictor in all the analyses of site-related macroinvertebrates biodiversity. This does not imply the abandonment of the BQIES in favor of the LHMS only because the latter was a better predictor of macroinvertebrates community assemblage. Simply, the different metrics represent different pressures with possible divergent scores of quality, and both have to be applied to highlight where to focus management or restoration actions more effectively.

**Supplementary Materials:** The following are available online at http://www.mdpi.com/2073-4441/12/9/2519/s1, Table S1: The Dataset with all the measurements for each single site within a lake, including lake name, site name, season, depth, temperature, oxygen, pH, conductivity, alkalinity, TP, TN, percentage of sand, percentage of silt, percentage of clay, water content, organic matter, carbonates, mineral content, richness of Chironomidae, richness of Oligochaeta, overall species richness, Shannon Diversity Index (SDI), and BQIES$_{single\text{-}site}$, LHMS, and LHQA. 'NA' means not available.

**Author Contributions:** Conceptualization, A.B. and D.F.; methodology, D.F.; software, D.F.; validation, A.B., S.Z., R.B., M.C. and D.F.; formal analysis, D.F.; investigation, A.B.; resources, A.B., S.Z. and M.C.; data curation, S.Z.; writing—original draft preparation, A.B., S.Z., R.B., M.C. and D.F.; writing—review and editing, A.B., S.Z., R.B., M.C. and D.F.; visualization, S.Z.; supervision, A.B.; project administration, A.B.; funding acquisition, M.C. All authors have read and agree to the published version of the manuscript.

**Funding:** This research was funded by EU Life + Project INHABIT (Local hydro-morphology, habitat and RBMPs: new measures to improve ecological quality in South European rivers and lakes—Contract No: LIFE08 ENV/IT/000413 INHABIT) under the LIFE + Environment Policy and Governance 2008 program.

**Conflicts of Interest:** The authors declare no conflict of interest.

## References

1. Geist, J. Integrative freshwater ecology and biodiversity conservation. *Ecol. Indic.* **2011**, *11*, 1507–1516. [CrossRef]

2. Geist, J.; Hawkins, S.J. Habitat recovery and restoration in aquatic ecosystems: Current progress and future challenges. *Aquat. Conserv Mar. Freshw. Ecosyst.* **2016**, *26*, 942–962. [CrossRef]

3. United Nations World Water Assessment Programme (WWAP). *The United Nations World Water Development Report 2018: Nature-Based Solutions*; UNESCO: Paris, France, 2018; p. 139.

4. Rouillard, J.; Lago, M.; Abhold, K.; Röschel, L.; Kafyeke, T.; Mattheiß, V.; Klimmek, H. Protecting aquatic biodiversity in Europe: How much do EU environmental policies support ecosystem-based management? *Ambio* **2018**, *47*, 15–24. [CrossRef]

5. Schneider, X. *D66: Best Practice Guidance for Water Body Monitoring*; XPRO Consulting Limited: Lefkosia, Cyprus, 2007; p. 36. Available online: https://www.xpro-consulting.com/uploads/4/9/5/5/49557869/d66-swift-best-practice.pdf (accessed on 29 June 2020).

6. Hering, D.; Verdonschot, P.F.M.; Moog, O.; Sandin, L. Integrated assessment of running waters in Europe. *Hydrobiologia* **2004**, *516*, 379.

7. Furse, M.; Hering, D.; Moog, O.; Verdonschot, P.; Johnson, R.K.; Brabec, K.; Gritzalis, K.; Buffagni, A.; Pinto, P.; Friberg, N.; et al. The STAR project: Context, objectives and approaches. *Hydrobiologia* **2006**, *566*, 3–29. [CrossRef]

8. Friberg, N.; Skriver, J.; Larsen, S.E.; Pedersen, M.L.; Buffagni, A. Stream macroinvertebrate occurrence along gradients in organic pollution and eutrophication. *Fresh. Biol.* **2010**, *55*, 1405–1419. [CrossRef]

9. Feld, C.K.; Borja, A.; Carvalho, L.; Hering, D.F. Water bodies in Europe: Integrative systems to assess ecological status and recovery—Results of the WISER project. *Hydrobiologia* **2013**, *704*, 501.

10. Taipale, S.J.; Vuorio, K.; Strandberg, U.; Kahilainen, K.K.; Jarvinen, M.; Hiltunen, M.; Peltomaa, E.; Kankaala, P. Lake eutrophication and brownification downgrade availability and transfer of essential fatty acids for human consumption. *Environ. Int.* **2016**, *96*, 156–166. [CrossRef]

11. Tolonen, K.T.; Vilmi, A.; Karjalainen, S.M.; Hellsten, S.; Heino, J. Do different facets of littoral macroinvertebrate diversity show congruent patterns in a large lake system? *Community Ecol.* **2017**, *18*, 109–116. [CrossRef]

12. Verdonschot, P.F. Integrated ecological assessment methods as a basis for sustainable catchment management. *Hydrobiologia* **2000**, *422*, 389–412. [CrossRef]

13. Birk, S.; Bonne, W.; Borja, A.; Brucet, S.; Courrat, A.; Poikane, S.; Solimini, A.G.; Bund, W. van de Zampoukas, N.; Hering, D. Three hundred ways to assess Europe's surface waters: An almost complete overview of biological methods to implement the water framework directive. *Ecol. Indic.* **2012**, *18*, 31–41. [CrossRef]

14. Bonada, N.; Prat, N.; Resh, V.; Statzner, B. Developments in aquatic insect biomonitoring: A comparative analysis of recent approaches. *Annu. Rev. Entomol.* **2006**, *51*, 495–523. [CrossRef] [PubMed]

15. Diaz, R.J.; Solan, M.; Valente, R.M. A review of approaches for classifying benthic habitats and evaluating habitat quality. *J. Environ. Manag.* **2004**, *73*, 165–181. [CrossRef] [PubMed]

16. Feio, M.J.; Poquet, J.M. Predictive models for freshwater biological assessment: Statistical approaches. Biological elements and the Iberian Peninsula experience: A review. *Int. Rev. Hydrobiol.* **2011**, *96*, 321–346. [CrossRef]

17. Lindegarth, M.; Carstensen, J.; Drakare, S.; Johnson, R.K.; Sandman, A.N.; Söderpalm, A.; Wikström, S.A. Ecological assessment of swedish water bodies; development, harmonisation and integration of biological indicators. In *Final Report of the Research Programme WATERS, Deliverable 1.1–4, WATERS Report No. 2016:10*; Havsmiljöinstitutet: Göteborg, Sweden, 2016. Available online: https://waters.gu.se/digitalAssets/1592/1592593_waters-report-2016_10.pdf (accessed on 29 June 2020).

18. McFarland, B.; Carse, F.; Sandin, L. Littoral macroinvertebrates as indicators of lake acidification within the UK. *Aquatic Conserv Mar. Freshw. Ecosyst.* **2010**, *20*, S105–S116. [CrossRef]

19. Poikane, S.; Birk, S.; Böhmer, J.; Carvalho, L.; de Hoyose, C.; Gassner, H.; Hellsten, S.; Kelly, M.; Solheim, A.L.; Olin, M.; et al. A hitchhiker's guide to European lake ecological assessment and intercalibration. *Ecol. Indic.* **2015**, *52*, 533–544. [CrossRef]

20. Reyjol, Y.; Argillier, C.; Bonne, W.; Borja, A.; Buijse, A.D.; Cardoso, A.C.; Daufresne, M.; Kernan, M.; Ferreira, M.T.; Poikane, S.; et al. Assessing the ecological status in the context of the European water framework directive: Where do we go now? *Sci. Total Environ.* **2014**, *497–498*, 332–344. [CrossRef]

21. Solheim, A.L.; Feld, C.K.; Birk, S.; Phillips, G.; Carvalho, L.; Morabito, G.; Mischke, U.; Willby, N.; Søndergaard, M.; Hellsten, S.; et al. Ecological status assessment of European lakes: A comparison of metrics for phytoplankton, macrophytes, benthic invertebrates and fish. *Hydrobiologia* **2013**, *704*, 57–74. [CrossRef]

22. Premazzi, G.; Dalmiglio, A.; Cardoso, A.; Chiaudani, G. Lake management in Italy: The implications of the Water Framework Directive. *Lakes Reserv. Res. Manag.* **2003**, *8*, 41–59. [CrossRef]

23. Rossaro, B.; Zaupa, S.; Lencioni, V.; Marziali, L.; Boggero, A. 6 Indice per la valutazione della qualità ecologica dei laghi italiani basato sulla comunità bentonica. In *Indici per La Valutazione Della Qualità Ecologica dei Laghi*; REPORT CNR ISE 02.13; Istituto Italiano di Idrobiologia: Verbania, Italy, 2013; pp. 93–113. Available online: http://www.ise.cnr.it/wfd (accessed on 29 June 2020).

24. Boggero, A.; Zaupa, S.; Cancellario, T.; Lencioni, V.; Marziali, L.; Rossaro, B. *Italian Classification Method for Macroinvertebrates in Lakes. Method Summary*; Report CNR ISE, 03.16; Istituto Italiano di Idrobiologia: Verbania, Italy, 2017; p. 16. Available online: http://www.ise.cnr.it/wfd-en (accessed on 29 June 2020).

25. Fornaroli, R.; Cabrini, R.; Zaupa, S.; Bettinetti, R.; Ciampittiello, M.; Boggero, A. Quantile regression analysis as a predictive tool for lake macroinvertebrate biodiversity. *Ecol. Indic.* **2016**, *61*, 728–738. [CrossRef]

26. Boggero, A.; Zaupa, S.; Rossaro, B.; Lencioni, V.; Marziali, L.; Buzzi, F.; Fiorenza, A.; Cason, M.; Giacomazzi, F.; Pozzi, S. Protocollo di Campionamento e Analisi dei Macroinvertebrati Negli Ambienti Lacustri. MATTM-APAT, Ed.; Roma, Italy, 2008. Available online: https://www.isprambiente.gov.it/files/pubblicazioni/manuali-lineeguida/metodi-biologici-acque/laghi-macroinvertebrati.pdf (accessed on 29 June 2020).

27. Carpenter, S.R.; Stanley, E.H.; Vander Zanden, M.J. State of the World's freshwater ecosystems: Physical, chemical, and biological changes. *Annu. Rev. Environ. Resour.* **2011**, *36*, 75–99. [CrossRef]

28. Ward, D.; Holmes, N.; José, P. *The New Rivers & Wildlife Handbook*; Royal Society for the Protection of Birds: Bedfordshire, UK, 1995.

29. Esteves, F.A. *Fundamentos de Limnologia*, 2nd ed.; Interciência/FINEP: Rio de Janeiro, Brazil, 1998.

30. Galdean, N.; Callisto, M.; Barbosa, F. Lotic Ecosystems of Serra do Cipó, southeast Brazil: Water quality and a tentative classification based on the benthic macro-invertebrate community. *Aquat. Ecosyst. Health* **2000**, *3*, 545–552. [CrossRef]

31. Meriläinen, J.J.; Hynynen, J.; Palomäki, A.; Mäntykoski, K.; Witick, A. Environmental history of an urban lake: A palaeolimnological study of Lake Jyväsjärvi Finland. *J. Paleolimnol.* **2003**, *30*, 387–406. [CrossRef]

32. Biggs, J.; von Fumetti, S.; Kelly-Quinn, M. The importance of small waterbodies for biodiversity and ecosystem services: Implications for policy makers. *Hydrobiologia* **2016**, *793*, 3–39. [CrossRef]

33. Buraschi, E.; Salerno, F.; Monguzzi, C.; Barbiero, G.; Tartari, G. Characterization of the Italian lake-types and identification of their reference sites using anthropogenic pressure factors. *J. Limnol.* **2005**, *64*, 75–84. [CrossRef]

34. Vollenweider, R.; Kerekes, J. *Eutrophication of Waters, Monitoring, Assessment and Control*; OECD: Paris, France, 1982; p. 154.

35. Tartari, G.A.; Mosello, R. *Metodologie Analitiche e Controlli di Qualità nel Laboratorio Chimico Dell'istituto Italiano di Idrobiologia*; Documenta; Consiglio Nazionale delle Ricerche: Roma, Italy, 1997; p. 160.

36. Rowan, J.S.; Carwardine, J.; Duck, R.W.; Bragg, O.M.; Black, A.R.; Cutler, M.E.J.; Soutar, I.; Boon, P.J. Development of a technique for lake habitat survey (LHS) with applications for the European Union Water Framework Directive. *Aquat. Conserv: Mar. Freshw. Ecosyst.* **2006**, *16*, 637–657. [CrossRef]

37. Dean, W.E. Determination of carbonate and organic matter in calcareous sediments and sedimentary rocks by loss on ignition: Comparison with other methods. *J. Sediment. Petrol.* **1974**, *44*, 242–248.

38. Wentworth, C.K. A scale of grade and class terms for clastic sediments. *J. Geol.* **1922**, *30*, 377–392. [CrossRef]

39. Ongley, E. Sediment measurements. In *Water Quality Monitoring—A Practical Guide to the Design and Implementation of Freshwater Quality Studies and Monitoring Programmes*; UNEP/WHO: London, UK, 1996; p. 15.

40. Sabra, N.; Dubourguier, H.C.; Hamieh, T. Sequential extraction and particle size analysis of heavy metals in sediments dredged from the Deule Canal, France. *Open Environ. Eng. J.* **2011**, *4*, 11–17. [CrossRef]

41. Wiederholm, T. Chironomidae of the Holartic region. Keys and diagnoses. Part I: Larvae. *Entomol. Scand. Suppl.* **1983**, *19*, 457.

42. Timm, T. A guide to the freshwater Oligochaeta and Polychaeta of Northern and Central Europe. *Lauterbornia* **2009**, *66*, 235.

43. Nocentini, A. *Guide per il Riconoscimento Delle Specie Animali Delle Acque Interne Italiane*; CNR: Verona, Italy, 1977–1985; Volume 29.

44. R Core Team. *R: A Language and Environment for Statistical Computing*; R Foundation for Statistical Computing: Vienna, Austria, 2019. Available online: https://www.R-project.org (accessed on 29 June 2020).

45. Revelle, W. *Psych: Procedures for Personality and Psychological Research*; Version 1.9.12; Northwestern University: Evanston, IL, USA, 2019. Available online: https://CRAN.R-project.org/package=psych (accessed on 29 June 2020).

46. Beckerman, A.P. What can modern statistical tools do for limnology. *J. Limnol.* **2014**, *73*, 161–170. [CrossRef]

47. Pinheiro, J.; Bates, D.; DebRoy, S.; Sarkar, D.; R Core Team. Nlme: Linear and Nonlinear Mixed Effects Models. R Package Version 3.1–147. 2020. Available online: https://CRAN.R-project.org/package=nlme (accessed on 29 June 2020).

48. Crawley, M.J. *The R Book*; John Wiley & Sons Ltd.: West Sussex, UK, 2007; p. 942.

49. Fox, J.; Weisberg, S. *An {R} Companion to Applied Regression*, 3rd ed.; Sage: Thousand Oaks, CA, USA, 2018.

50. Jaeger, B. r2glmm: Computes R Squared for Mixed (Multilevel) Models. R Package Version 0.1.2. 2017. Available online: https://CRAN.R-project.org/package=r2glmm (accessed on 29 June 2020).

51. Mac Donald, D.D.; Ingersoll, C.G.; Berger, T.A. Development and evaluation of consensus-based sediment quality guidelines for freshwater systems. *Arch. Environ. Contam. Toxicol.* **2000**, *39*, 20–31. [CrossRef] [PubMed]

52. Bettinetti, R.; Zaupa, S.; Fontaneto, D.; Boggero, A. Biological, chemical, and ecotoxicological assessments using benthos provide different and complementary measures of lake ecological status. *Water* **2020**, *12*, 1140. [CrossRef]

53. Sarukhán, J.; Whyte, A.; Hassan, R.; Scholes, R.; Ash, N.; Carpenter, S.T.; Leemans, R. Millenium ecosystem assessment: Ecosystems and human well-being. In *The Millennium Ecosystem Assessment Series*; Hassan, R., Scholes, R., Neville, A., Eds.; Island Press: Washington, DC, USA, 2003; p. 245.

54. Free, G.; Solimini, A.; Rossaro, B.; Marziali, L.; Giacchini, R.; Paracchini, P.; Ghiani, M.; Vaccaro, S.; Gawlik, B.; Fresner, R.; et al. Modelling lake macroinvertebrate species in the shallow sublittoral, relative roles of habitat, lake morphology, aquatic chemistry and sediment composition. *Hydrobiologia* **2009**, *133*, 123–136. [CrossRef]

55. Garcia-Criado, F.; Becares, E.; Fernadez-Alaez, M. Plant-associated invertebrates and ecological quality insome Mediterranean shallow lakes: Implications for the application of the EC Water Framework Directive. *Aquat. Conserv: Mar. Freshwat. Ecosyst.* **2005**, *15*, 31–50. [CrossRef]

56. Rossaro, B.; Boggero, A.; Lencioni, V.; Marziali, L.; Solimini, A. Tools for the development of a benthic quality index for Italian lakes. *J. Limnol.* **2006**, *65*, 41–51. [CrossRef]

57. Rossaro, B.; Cardoso, A.C.; Solimini, A.; Free, G.; Marziali, L.; Giacchini, R. A biotic index using benthic macroinvertebrates for Italian lakes. *Ecol. Indic.* **2007**, *7*, 412–429. [CrossRef]

58. Donohue, I.; Garcia Molinos, J. Impacts of increased sediment loads on the ecology of lakes. *Biol. Rev.* **2009**, *84*, 517–531. [CrossRef]

59. Dias-Silva, K.; Cabette, H.S.R.; Juen, L.; DeMarco, P. The influence of habitat integrity and physical-chemical water variables on the structure of aquatic and semi-aquatic Heteroptera. *Zoologia* **2010**, *27*, 918–930. [CrossRef]

60. Jenny, H. *The Soil Resource*, 1st ed.; Springer: New York, NY, USA, 2008.

61. Stohlgren, T.; Bachand, R.; Onami, Y.; Binkley, D. Species-environment relationships and vegetation patterns: Effects of spatial scale and tree life-stage. *Plant Ecol.* **1998**, *135*, 215–228. [CrossRef]

62. Hynes, H.B.N. The stream and its valley. *Verh. Internat. Verein. Theor. Angew. Limnol.* **1975**, *19*, 1–15. [CrossRef]

63. Larouche, J.R.; Bowden, W.B.; Giordano, R.; Flinn, M.B.; Crump, B.C. Microbial biogeography of Arctic streams: Exploring influences of lithology and habitat. *Front. Microbiol.* **2012**, *3*, 309. [CrossRef] [PubMed]

64. McGoff, E.; Aroviita, J.; Pilotto, F.; Miler, O.; Solimini, A.G.; Porst, G.; Jurca, T.; Donohue, L.; Sandin, L. Assessing the relationship between the lake habitat survey and littoral macroinvertebrate communities in European lakes. *Ecol. Ind.* **2013**, *25*, 205–214. [CrossRef]

*Article*

# How to Assess the Ecological Status of Highly Humic Lakes? Development of a New Method Based on Benthic Invertebrates

Dāvis Ozoliņš [1,*], Agnija Skuja [1], Jolanta Jēkabsone [1], Ilga Kokorite [1], Andris Avotins [2,3] and Sandra Poikane [4,*]

1    Laboratory of Hydrobiology, Institute of Biology, University of Latvia, Jelgavas 1, LV-1004 Riga, Latvia; agnija.skuja@lu.lv (A.S.); jolanta.jekabsone@lvgmc.lv (J.J.); ilga.kokorite@lu.lv (I.K.)
2    Statistics Unit, Riga Stradiņš University, Baložu 14, LV-1048 Riga, Latvia; andris.avotins@rsu.lv
3    Institute of Food Safety, Animal Health and Environment BIOR, Lejupes 3, LV-1076 Riga, Latvia
4    European Commission, Joint Research Centre, 21027 Ispra, Italy
*    Correspondence: davis.ozolins@lu.lv (D.O.); sandra.poikane@ec.europa.eu (S.P.); Tel.: +371-29390452 (D.O.)

**Abstract:** Highly humic lakes are typical for the boreal zone. These unique ecosystems are characterised as relatively undisturbed habitats with brown water, high acidity, low nutrient content and lack of macrophytes. Current lake assessment methods are not appropriate for ecological assessment of highly humic lakes because of their unique properties and differing human pressures acting on these ecosystems. This study proposes a new approach suitable for the ecological status assessment of highly humic lakes impacted by hydrological modifications. Altogether, 52 macroinvertebrate samples from 15 raised bog lakes were used to develop the method. The studied lakes are located in the raised bogs at the central and eastern parts of Latvia. Altered water level was found as the main threat to the humic lake habitats since no other pressures were established. A multimetric index based on macroinvertebrate abundance, littoral and profundal preferences, Coleoptera taxa richness and the Biological Monitoring Working Party (BMWP) Score is suggested as the most suitable tool to assess the ecological quality of the highly humic lakes.

**Keywords:** highly humic lakes; macroinvertebrates; ecological status assessment



**Citation:** Ozoliņš, D.; Skuja, A.; Jēkabsone, J.; Kokorite, I.; Avotins, A.; Poikane, S. How to Assess the Ecological Status of Highly Humic Lakes? Development of a New Method Based on Benthic Invertebrates. *Water* **2021**, *13*, 223. https://doi.org/10.3390/w13020223

Received: 7 December 2020
Accepted: 14 January 2021
Published: 18 January 2021

**Publisher's Note:** MDPI stays neutral with regard to jurisdictional claims in published maps and institutional affiliations.



## 1. Introduction

Humic lakes, also known as brown-water lakes, are typical for the boreal zone, located between 50° to 70° N latitude. These ecosystems are characterized by dark water colour, low water transparency, and low pH caused by the high concentration of dissolved organic matter (DOM), mostly originating from the catchment area and consisting of refractory humic substances [1,2]. Over the last decades, an impressive body of evidence has been accumulated which suggests that DOM is a major modulator of the structure and function of lake ecosystems, affecting numerous features such as light regime, thermal stratification, nutrient availability, primary production, and microbial metabolism [3–6]. Numerous studies have shown that the biological communities differ considerably from those of clear-water lakes (phytoplankton: [7]; macrophytes: [8]; periphyton: [9]; zooplankton: [10]; fish fauna: [11]). Moreover, their response to human stressors might differ too [12–14], asking for monitoring and assessment approaches different to those used for clear-water lakes [15,16].

Multiple human pressures, such as nutrient enrichment, hydrological and morphological alterations, invasion of non-native species and climate change, affect humic lakes [17,18]. Some of these pressures are similar to those impacting clear water lakes, but some are different, e.g., bog lakes are affected by artificial peatland drainage and peat extraction associated with habitat degradation, erosion, increased leaching of nutrients and dissolved organic carbon [19,20]. Similarly, peatland lakes are impacted by forestry practices (afforestation, fertilization, and clear-cutting) which have been shown to increase catchment loadings of nutrients, sediments and dissolved organic carbon (DOC) to these ecosystems [21–23]. These impacts can have a profound effect on the water quality, lake habitats, associated biological assemblages and conservation value of these lakes [22,24,25].

In Europe, the Water Framework Directive (WFD) [26] establishes a framework for the protection of inland and coastal waters. According to the WFD, lakes have to be classified into five status classes (high, good, moderate, poor and bad) based on four biological quality elements (BQEs)—phytoplankton, benthic flora, benthic invertebrates, and fish fauna. In addition, physico-chemical elements (e.g., nutrient conditions) and hydromorphological elements (e.g., flow conditions) are used to support ecological classification. The European Union (EU) member states have to identify degraded water bodies (i.e., less than good status) and to establish programmes of measures for each river basin district to reduce significant anthropogenic pressures and achieve good water status.

A large number of lake assessment methods have been intercalibrated and included in the European countries' monitoring toolkits [27]. In the recent years, all of these methods have been intercalibrated (i.e., compared and harmonized) among the EU member states [28,29]. Lake assessment methods include both primary producers, e.g., phytoplankton (e.g., [30,31]), macrophytes (e.g., [32,33]), phytobenthos (e.g., [34]), and heterotrophs such as benthic invertebrates and fish fauna (e.g., [33,35,36]).

However, two problems are still overlooked. At first, the majority of assessment systems target nutrient enrichment, while other key pressures are largely neglected. This is especially true regarding hydromorphological pressures, which affect a considerable number of lakes across Europe [37]. Only few assessment systems tackle the ecological effects of these pressures (shore degradation: [38]; water level fluctuations: [39]) and only two of these systems have been intercalibrated among member states [36].

Second, despite the well-known differences among clear and brown-water lakes [1], the current lake assessment systems are adopted mostly for clear-water lakes. Recently, several studies have raised the issue that assessment systems might not be appropriate for humic lake assessment [13,17,40]. Hence, there is an imperative need for the development of appropriate assessment methods targeting humic lakes.

Studies on macroinvertebrates in highly humic lakes are mainly focused either on biodiversity [41,42] or specific taxonomic groups, e.g., chironomids [43] or Coleoptera [44]. Raised bog water bodies are also known as habitats for rare and protected macroinvertebrate species [41].

Mires and bogs cover 4.9% of the territory of Latvia [45] ranking Latvia number 9 by the total area of peatlands among all the European countries [46]. Bog lakes are listed as protected habitats within the EU Habitats Directive emphasizing their high conservation value [47]. However, these lakes have been impacted by a range of anthropogenic activities, most importantly anthropogenic drainage and peat harvesting which can lead to water level fluctuations, loss of biodiversity and degradation of the lake ecological status [48,49]. Nevertheless, the effects of these impacts are poorly understood and there are no assessment tools in place to assess the ecological condition of humic lakes. According to the River Basin Management Plans (RBMPs) of Latvia, current methods reflect bog lakes at poor or bad status, though the anthropogenic pressures are irrelevant [50]. As a result, there is a need to develop new methods for the ecological assessment of the highly humic lakes.

The objectives of this study are (i) to explore littoral benthic invertebrate community response to hydrological alterations in highly humic lakes; (ii) to develop a biotic index for assessing hydrological alterations in these aquatic ecosystems.

## 2. Materials and Methods

### 2.1. Study Sites

Altogether 15 highly humic lakes were studied at seven raised bogs comprising the national monitoring data from the Latvian Environmental, Geology and Meteorology Centre and studies from the Institute of Biology, University of Latvia. The lakes were divided into two groups: lakes with altered water level due to drainage and lakes with natural or restored hydrological regime (see Figure 1).

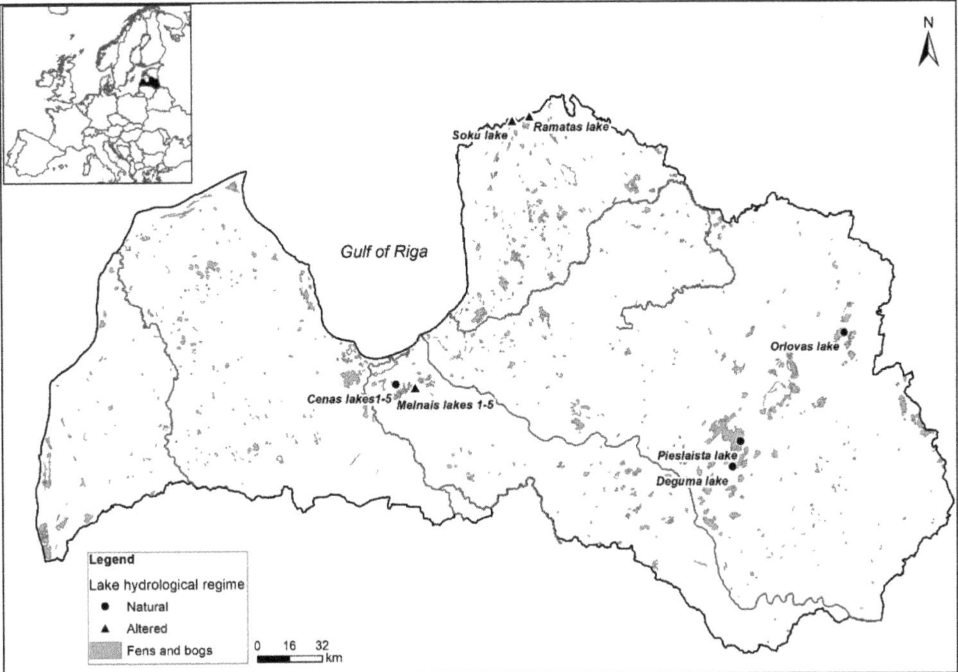



**Figure 1.** The macroinvertebrate sampling sites and distribution of fens in Latvia. Black triangles: highly humic lakes with altered water level; black dots: highly humic lakes with natural or restored water level.

All of the studied lakes are located in the territories included in the Natura 2000 network. In the central part of Latvia, Cena Mire and Melnais Lake Mire are represented by the samples from five small bog lakes each. In the Cena Mire, lakes are located in pristine areas, while in the Melnais Lake Mire—close to the peat excavation fields, thus representing a hydrologically disturbed state. In the northern part of Latvia, sampling was conducted in Lake Ramatas Lielezers and Lake Soku, both being with altered water level due to outflowing drainage ditches. The water level was receded at these lakes with visible open peat outcrops at the shoreline (see Figure 2).

Eastern part of Latvia is represented by three lakes (Pieslaista, Deguma and Orlovas) that have an unaltered or restored hydrological regime. The water level at these lakes was natural and not affected due to the drainage ditches or peat excavation (see Figure 3).

**Figure 2.** Lake Soku with receded water level and visible open peat outcrops at the shoreline in May 2019.

**Figure 3.** Lake Deguma with natural water level and mire vegetation at the shoreline in May 2019.

*2.2. Physical and Chemical Parameters*

Waterbodies of Melnais Lake Mire and Cena Mire were sampled in May 2015 and analyses were conducted at the Laboratory of Soils of the University of Latvia. Soku, Ramatas Lielezers, Deguma, Pieslaista and Orlovas lakes were sampled four times in different seasons in 2017 and analysis was conducted at the Laboratory of Latvian Environment, Geology and Meteorology Centre. In 2015, pH and electric conductivity (EC) were measured in-situ by using a portable pH tester (HI 98127, HANNA instruments, Sarmeola di Rubano, Italy) and conductivity tester (The Original Dist HI 98300, HANNA instruments, Sarmeola di Rubano, Italy). In 2017, these parameters were measured in-situ by using a portable probe (HQ40d, Hach Companies, Loveland, CO, USA). Total phosphorus (TP) was detected by ascorbic acid method after the digestion using potassium persulfate. Total nitrogen (TN) samples were digested by potassium persulfate, then nitrates were reduced to nitrites in a Cd column and analysed spectrophotometrically. Water colour was analysed spectrophotometrically using the Pt/Co scale [51].

### 2.3. Macroinvertebrate Sampling and Sample Processing

In larger lakes, sampling was conducted in a 100 m long, representative shoreline section, while in smaller lakes sampling was conducted around all of the shoreline. Benthic macroinvertebrate samples were taken by hand net (frame size 0.25 × 0.25 m, mesh of 0.5 mm) using the sweeping technique. A hand net was placed on the bottom parallel to the shore, if the depth was less than 0.8 m or under the vegetation overhang at the same depth and moved upwards over the vegetation stands to the surface. Five replicates of 0.5 m sweeps were chosen in proportion to habitat types, e.g., of stands of *Menyanthes trifoliata*, *Sphagnum cuspidatum*, *Batrachospermum turfosum*, *Carex* spp., bare littoral, etc. At the small lakes of Melnais Lake Mire and Cena Mire, samples were taken in May 2015. Sampling at Soku, Ramatas Lielezers, Deguma, Pieslaista and Orlovas lakes was conducted in May and October 2017. Additional samples were taken at lakes Soku, Ramatas Lielezers and Deguma in May 2019.

Sampled material was washed through a sieve with a mesh size of 0.5 mm at field. All replicates were placed in the polyethylene containers, labelled and preserved in 70% ethyl alcohol (final concentration).

Preserved samples were washed at the laboratory; all specimens were picked out from the vegetation, detritus and peat particles. Macroinvertebrates were identified to the smallest achievable taxonomical (species, genera) level, excluding Oligochaeta and juvenile Hydrachnidia. Specimens of Diptera were identified to the family level.

### 2.4. Data Analysis

#### 2.4.1. Selection of Metrics

According to the WFD, the ecological quality assessment indices are required multimetric consisting of composition, abundance, sensitive/tolerant taxa and diversity metrics [26]. Macroinvertebrate metrics were calculated using ASTERICS 4.0.4. software (Wageningen Software Labs, Wageningen, The Netherlands) [52]. Numerically unsuitable metrics and majority of the metrics specific for the lotic habitats were excluded from further analysis. In total, 139 indices describing 52 samples from water bodies in open raised bogs were tested for a selection of multimetric index according to requirements of the EU WFD [26]. We generally followed the procedure described by Hering et al. [53], beginning with the reduction of dimensionality by the evaluation of each metric value distribution between the altered and the natural water bodies.

#### 2.4.2. Sensitivity to Stressor

Only descriptors correlating to the stressor gradient can be used in the development of a multimetric index. We used Mann–Whitney–Wilcoxon U tests (each metric as dependent in groups of stressor, thereinafter U test) together with boxplots and simple binary logistic regression (groups of stressor as dependant: altered = 0, natural = 1, thereinafter binary logistic regression (BLR)) to select only statistically significant metrics for further processing. Metric values prior to BLR were centred by mean and scaled by standard deviation to ease the convergence, while unscaled values were used in the boxplots and U test.

With type I statistical error rate 0.05, the U test returned 66 statistically significant indices and BLR-44 (Table S1). All the indices found significant by BLR were significant also by U test. As a reason for this step is the reduction of dimensionality with selection of potentially most important indices, we did not account for possible false discoveries due to the multiple testing. We included metrics (except life index due to specific relation to lotic environment) found significant by both methods in further investigation of their suitability.

### 2.4.3. Numerical Suitability

The multimetric index must consist of metrics that tend to describe large gradients with possibly low skew and preferably without outliers [53]. These numeric properties can be the best evaluated graphically if values are from the same or a similar scale. We used the violin (density) plots with a point overlay (with stressor group indicated by a shape) to select one to six metrics per a metric type. Before plotting, each metric was scaled by its observed maximum value. Graphs used in evaluation are provided in Figure 4 with 14 metrics included in the further investigation marked with the dark background.

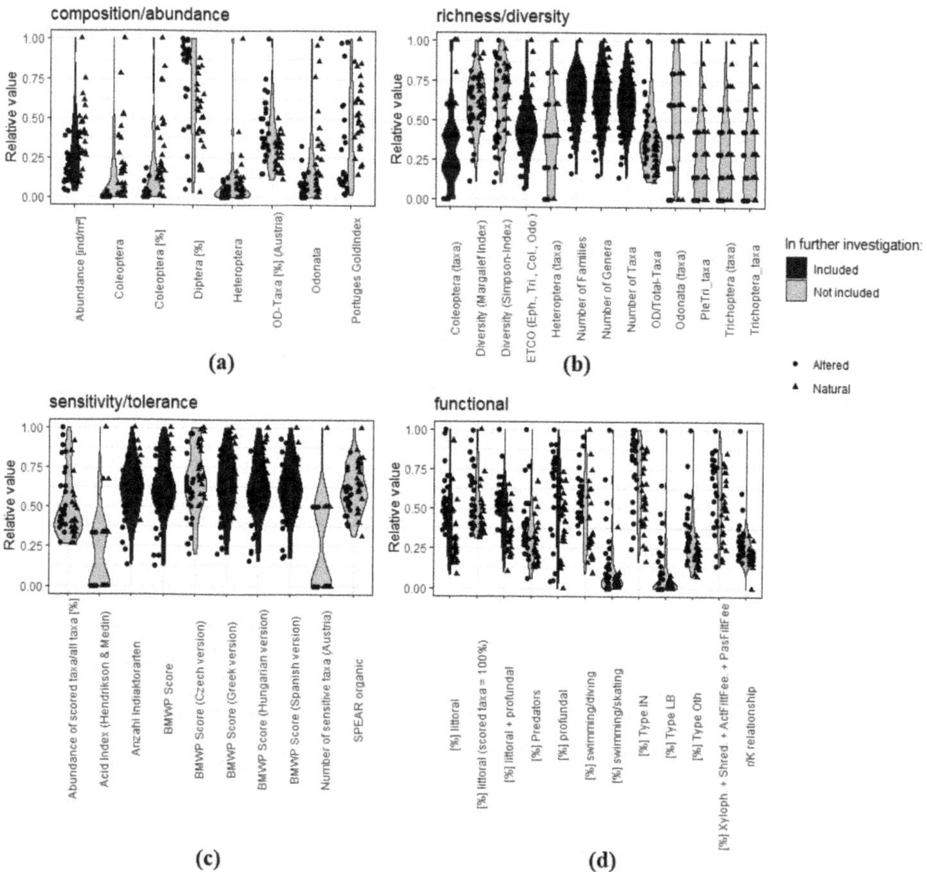

**Figure 4.** The evaluation graphs of (**a**) composition/abundance; (**b**) richness/diversity; (**c**) sensitive/tolerance and (**d**) functional metric groups of highly humic lakes.

### 2.4.4. Ecological Relevance

The group of experts authoring this paper evaluated each previously selected metric to exclude possibly non-explanatory correlations and those of ecologically low meaning. Additionally, metrics created for ecoregions other than boreal were excluded resulting in ten metrics for further use.

### 2.4.5. Correlations

Metrics with Spearman's correlation coefficient |≥0.8| are considered redundant and only one of them must be used [53]. Additionally, we tried to avoid possible multicollinearity by selecting metrics with even lower (|≤0.6|) values [54] to obtain at least one metric per one WFD criteria. Correlation coefficients of the selected metrics are available in Table S2.

At this point, we decided to keep following metrics:

- For composition/abundance: abundance [ind/m$^2$];
- For richness/diversity: Coleoptera (taxa);
- For sensitivity/tolerance: BMWP (Biological Monitoring Working Party) Score;
- For functional metrics: (%) littoral, (%) littoral + profundal, (%) profundal.

As the selected functional metrics are highly correlated both statistically and ecologically, we decided to compare three models with inclusion of every previously mentioned metric from above and functional metrics as follows:

- (%) littoral;
- (%) littoral + profundal;
- (%) profundal.

As in every principal model where metric types have the same weights, we used the mean value as the result for the multimetric index.

### 2.4.6. Scaling

To ensure that the multimetric index is limited between 0 and 1, and to avoid potential influence of some extremely high- or low-quality sites anchoring is suggested [53]. We used the same approach for metrics decreasing with increasing impairment, but corrected the approach for the metrics increasing with increasing impairment to:

Value = 1—(Metric result—Lower Anchor)/(Upper Anchor—Lower Anchor). We corrected values >1 to 1 and negative values to 0.

To ensure good spread of sites within multimetric index, we compared several anchor values (Table S3):

- 5th percentile and 95th percentile;
- 10th percentile and 90th percentile;
- 10th percentile and 80th percentile;
- Each of the previous with prespecified values for "Coleoptera (taxa)" as 0 for lower and 4 for upper.

### 2.4.7. Quality Classification

We used the quality classes in accordance with WFD demands, following suggestion of Hering et al. [53]:

- Reference ≥0.8;
- Good ≥0.6 < 0.8;
- Moderate ≥0.4 < 0.6;
- Poor ≥0.2 < 0.4;
- Bad <0.2.

We consider an index to be the best, if natural lakes are concentrated at the reference and good quality classes, while altered lakes are at bad- and poor-quality classes with some mixture present in a class of moderate quality.

We used the software R 4.0.3 (The R Foundation for Statistical Computing, Vienna, Austria) for data analysis [55]. Data processing and visualisations were performed within the tidyverse ecosystem [56].

## 3. Results

### 3.1. Characterisation of Chemical and Environmental Variables

The studied lakes are poly-humic lakes as indicated by high water colour values (114–666 mg Pt/L) and low pH values (3.35–6.09). Electric Conductivity (EC) is in the range of 21–65 µS/cm, concentrations of TN are 0.43–1.68 mg/L and TP are 0.017–0.061 mg/L. The highest conductivity, water color and total nitrogen values were observed in waterbodies of the Melnais Lake Mire (see Table 1).

**Table 1.** Mean annual chemical and environmental parameters at the studied highly humic lakes (Cond—electric conductivity, TN—total nitrogen, TP—total phosphorus).

| Lakes | Altered Water Level (+) | Year | pH | Cond µS/cm | Colour mg Pt/L | TN mg/L | TP mg/L |
|---|---|---|---|---|---|---|---|
| Cenas Mire Lake 1 | | 2015 | 3.93 | 28 | 124 | 0.95 | 0.017 |
| Cenas Mire Lake 2 | | 2015 | 4.44 | 29 | 144 | 0.99 | 0.020 |
| Cenas Mire Lake 3 | | 2015 | 4.49 | 26 | 114 | 0.90 | 0.019 |
| Cenas Mire Lake 4 | | 2015 | 3.75 | 32 | 189 | 0.83 | 0.021 |
| Cenas Mire Lake 5 | | 2015 | 3.46 | 43 | 304 | 0.92 | 0.020 |
| Melnais Lake Mire Lake 1 | + | 2015 | 3.68 | 65 | 393 | 1.31 | 0.019 |
| Melnais Lake Mire Lake 2 | + | 2015 | 3.53 | 49 | 402 | 1.25 | 0.022 |
| Melnais Lake Mire Lake 3 | + | 2015 | 3.58 | 44 | 365 | 1.17 | 0.022 |
| Melnais Lake Mire Lake 4 | + | 2015 | 3.43 | 45 | 505 | 1.36 | 0.017 |
| Melnais Lake Mire Lake 5 | + | 2015 | 3.35 | 48 | 666 | 1.68 | 0.028 |
| Deguma Lake | | 2017 | 5.09 | 31 | 222 | 0.95 | 0.032 |
| Orlovas Lake | | 2017 | 5.42 | 21 | 205 | 0.87 | 0.037 |
| Pieslaista Lake | | 2017 | 5.02 | 36 | 238 | 0.65 | 0.061 |
| Ramatas Lielezers Lake | + | 2017 | 6.09 | 23 | 134 | 0.65 | 0.032 |
| Soku Lake | + | 2017 | 5.88 | 25 | 130 | 0.43 | 0.035 |

### 3.2. Benthic Invertebrate Taxa

A list of benthic invertebrate taxa found in highly humic lakes is presented in the Supplemental Material (Table S4). Altogether, 18,808 individuals from 106 macroinvertebrate taxa are recorded at the studied bog lakes, of which the orders Coleoptera, Odonata and Trichoptera have the highest species richness. Mayflies (Ephemeroptera) are represented by four species, of which *Leptobphlebia vespertina* is the most widespread, missing only in the water bodies of the Cena Mire. In the humic lakes of Melnais Lake Mire and Cena Mire, larvae of dragonflies *Leucorrhinia albifrons* and *L. pectoralis* are recorded. These two species are included in the Bern Convention [57] and in the Habitats Directive [47]. The highest abundance of macroinvertebrates is recorded from waterbodies of Melnais Lake Mire and Cena Mire, ranging from 813 to 2014 individuals per sample, while the benthic invertebrate abundance at larger lakes ranges from 51 to 1202 individuals.

The macroinvertebrate orders Diptera and Coleoptera are the most abundant taxa at the lakes of the Cena Mire, while chironomids dominate in lakes of the Melnais Lake Mire. Taxonomic structure at lakes Orlovas, Deguma, Pieslaista, Ramatas Lielezers and Soku vary due to repeated sampling in different years and seasons, generally with Diptera, Coleoptera and Ephemeroptera as the dominant taxa. Molluscs (Gastropoda and Bivalvia) are completely absent at the studied lakes (see Figure 5).

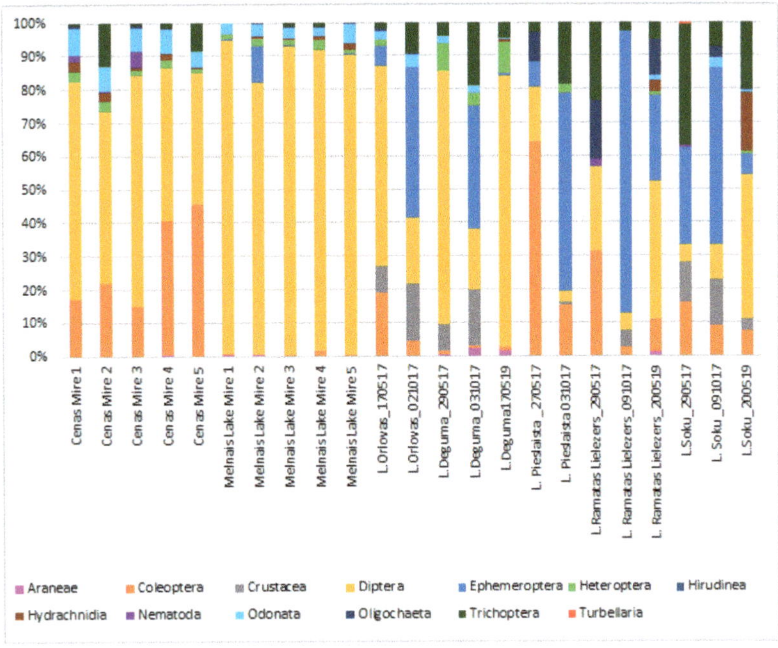

**Figure 5.** The dominance structure of macroinvertebrate assemblages in highly humic lakes of Latvia.

### 3.3. Metrics

The comparison of macroinvertebrate abundance (ind/m$^2$) between samples of bog lakes with altered water level and those with natural water level show significant differences ($p < 0.001$) (Figure 6). The macroinvertebrate abundance in lakes with natural level varies from 109 to 1202 individuals, while in lakes with altered level from 51 to 502. Similar significant differences are found between the number of taxa varying from 9 to 22 taxa in natural lakes and 3 to 18 taxa in altered lakes ($p < 0.001$). Additionally, our results show higher values of BMWP Score at natural lakes rather than the altered ones ($p = 0.002$). The BMWP values in natural lakes vary from 33 to 74, whereas from 10 to 67 in bog lakes with altered water level. Taxa richness of Coleoptera and ETCO (Ephemeroptera, Trichoptera, Coleoptera, Odonata) show the same significant differences between the studied lakes ($p < 0.001$). Natural bog lakes are represented by 1 to 5 Coleoptera species, while in the altered lakes 0 to 3 species are found. The number of ETCO varies from 5 to 14 species in natural lakes while 1 to 9 species in bog lakes with altered water level, respectively. Nevertheless, altered lakes are represented by taxa preferring littoral and profundal habitats ($p < 0.001$) and higher percentage of Diptera ($p < 0.001$). Number of taxa of Trichoptera ($p = 0.023$), Odonata ($p = 0.009$) and Heteroptera ($p = 0.007$) is higher at humic lakes with natural water level.

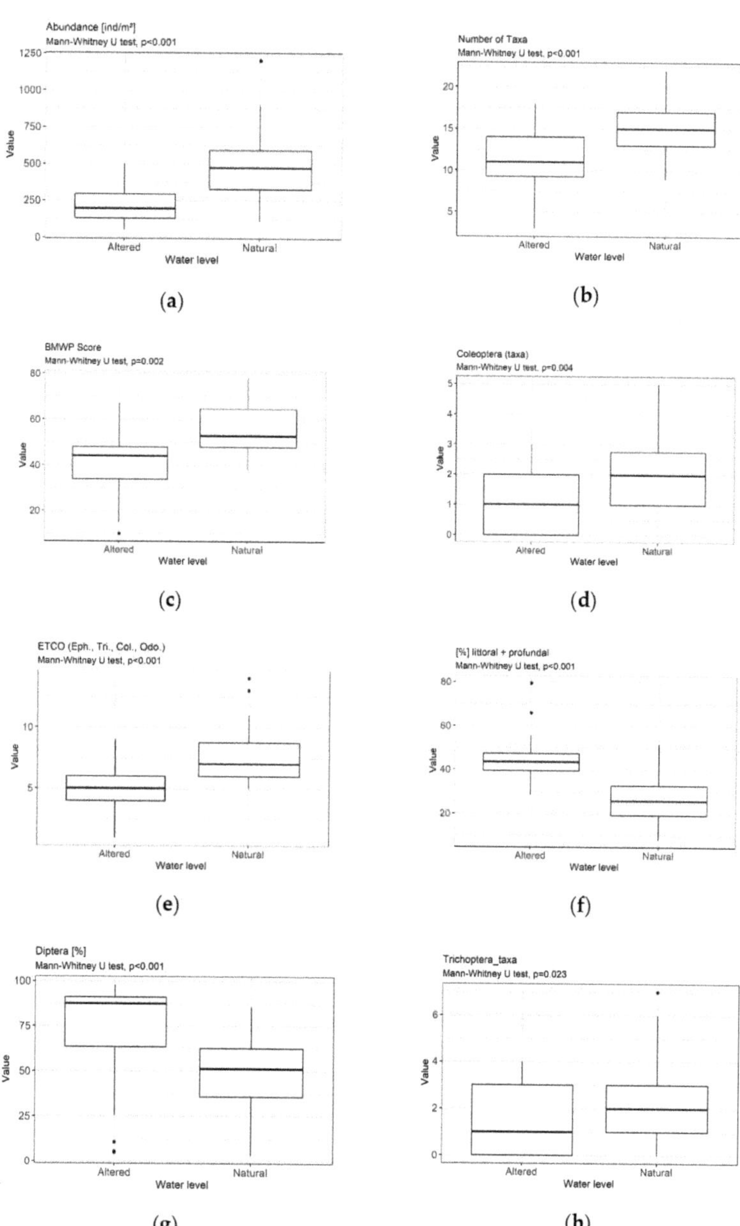

**Figure 6.** *Cont.*

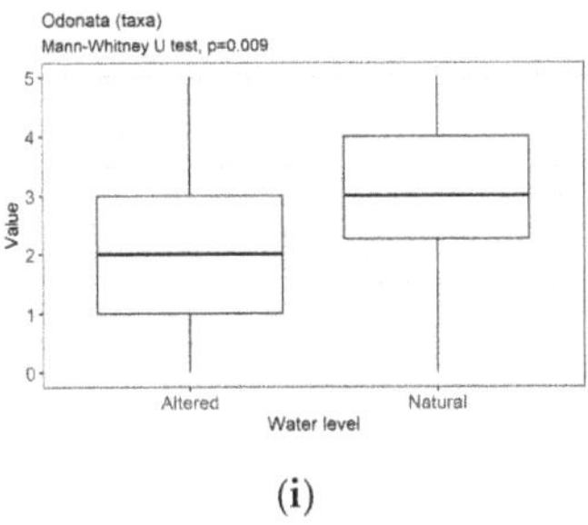

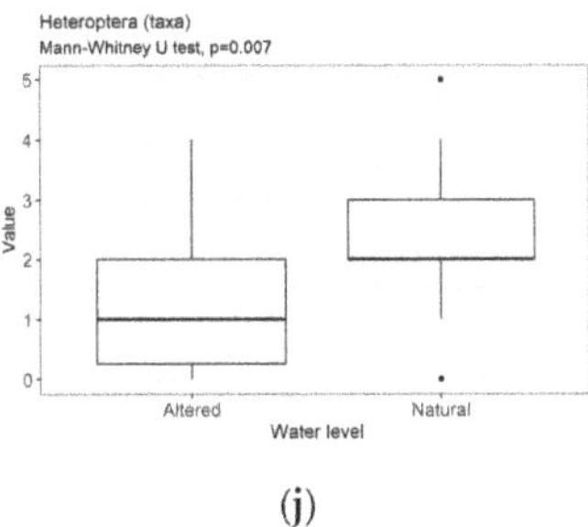

(i)           (j)

**Figure 6.** Differences of macroinvertebrate (**a**) abundance, (**b**) taxa richness, (**c**) BMWP (Biological Monitoring Working Party) Score, (**d**) Coleoptera taxa, (**e**) ETCO (Ephemeroptera, Trichoptera, Coleoptera, Odonata), (**f**) littoral + profundal preference, (**g**) % Diptera, (**h**) Trichoptera taxa, (**i**) Odonata taxa and (**j**) Heteroptera taxa between bog lakes with altered and natural level.

### 3.4. Multimetric Index

We found the model 2 (Figure 7) ((%) littoral + profundal and abundance (ind/m$^2$) and Coleoptera (taxa) and BMWP Score) with P10 and P80 anchoring and predefined Coleoptera (taxa) anchoring to be the best classifier for the highly humic lakes.

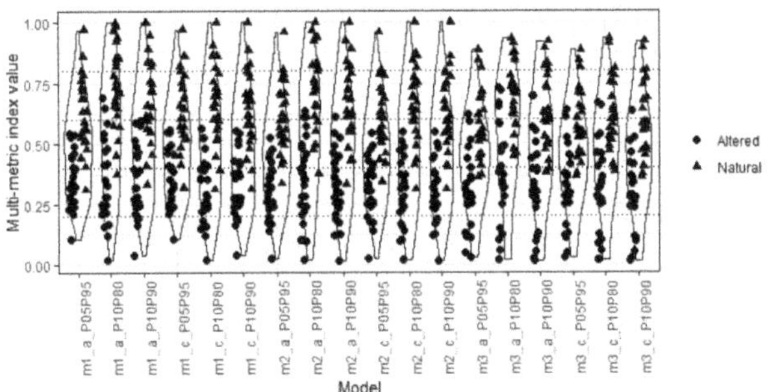

**Figure 7.** The model comparison of the humic lake multimetric indices at the lakes with natural and altered water level. The model caption consists of model_Coleoptera (c = yes; a = no) percentile anchors.

The distribution of values of the highly humic lake multimetric index are shown in Figure 8.

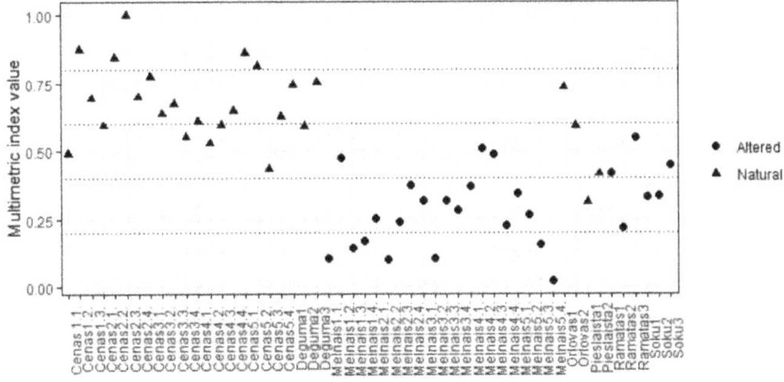

**Figure 8.** The distribution of values of the newly developed highly humic lake multimetric index at the lakes with natural and altered water level.

## 4. Discussion

The total number of taxa found in the studied highly humic lakes is relatively low and similar to those of other studies [40,41]. Desrochers and van Duinen [58] noted that the chemical constraints of humic lakes almost entirely exclude several macroinvertebrate taxa, e.g., lumbricid worms, isopods, and snails and the low nutrient availability may limit the presence of species with high nutrient preferences. In addition, acidity is known as a limiting factor for many macroinvertebrate groups, e.g., freshwater snails [59], mussels [60] and mayflies [61]. In general, macroinvertebrate species specialized on different mire habitats are obviously able to realize their life cycle at water pH of 4.0–5.0 [62]. The macroinvertebarte taxonomical composition and lack of molluscs indicate the peculiarities of the studied highly humic lakes.

The invertebrate assemblage in altered lakes showed a reduced taxonomic richness, especially in Trichoptera, Coleoptera, Heteroptera and Odonata, an increased proportion of Diptera and an overall lower numerical abundance.

The best-performing metrics were total abundance (for composition/abundance category), number of Coleoptera taxa (for richness/diversity category), BMWP score (for sensitivity/tolerance category), and % littoral and profundal (for functional metric category).

### 4.1. Use of Benthic Invertebrates in Lake Ecological Assessment

In lakes, phytoplankton and macrophytes are the most widely used communities for ecological assessment. However, in the last decades numerous systems using benthic invertebrates have been developed, following the requirements of the WFD [36]. These systems differ by habitat sampled (mostly littoral, but some sampled the profundal), putative pressure assessed and metrics used.

According to the concept of multimetric approach [53,63] different types of metrics should be included into the assessment system (composition/abundance metrics, richness/diversity metrics; sensitivity/tolerance metrics; functional metrics).

### 4.1.1. Sensitivity/Tolerance Metrics

Almost all lake benthic invertebrate assessment systems include some kind of sensitivity index, most common are Average Score per Taxon (ASPT) index [64], Acid Water Indicator Community (AWIC) index [65], Benthic Quality Index [66] and Fauna index [67].

Originally, ASPT and BMWP indices have been developed for river water quality in Britain [68], however they have proved suitable for lake assessment. For instance, Šidagytė et al. [69] have demonstrated relationships between ASPT and total phosphorus, biochemical oxygen demand (BOD) and tropho-morphoindex in lakes of Lithuania, while Mavromati et al. [70]—relationships between ASPT and the total phosphorus and shoreline modification in lakes for Greece. Similarly, relationships between ASPT and eutrophication and other pressures have been demonstrated in lakes of Denmark [71]. Currently, ASPT index is used in the lake assessment in Denmark, Greece, Estonia, Latvia, Lithuania, Poland, and Sweden [36] while BMWP in Hungary and Bulgaria [72]. Additionally, in our study, we found that ASPT and BMWP differentiate between natural and altered humic lakes, though we chose to include BMWP ($p = 0.002$) rather than ASPT ($p = 0.45$).

### 4.1.2. Richness/Diversity Metrics

Richness diversity metrics are the widely represented metric category in the lake assessment: almost all the countries use some of these metrics: total taxa richness, Shannon diversity, Margalef diversity or other. Many studies have revealed relationships between total taxa richness/diversity metrics and different human pressures: morphological degradation [73], water-level fluctuation [74], total phosphorus [70] and integrated pressure index [71]. However, in many cases, a pooled taxa number of stressor-sensitive macroinvertebrate orders can be more informative: for instance, CEP taxa richness (Coleoptera, Ephemeroptera, Plecoptera) has shown relatively strong relationships with a range of eutrophication indicators and hydromorphological index in Lithuania, so it was included in the Lithuanian assessment system [69]. Similarly, number of EPTCBO (Ephemeroptera, Plecoptera, Trichoptera, Coleoptera, Bivalvia, Odonata) taxa is used in Denmark, based on demonstrated relationships with eutrophication pressure [71]. In contrast, we found a significant difference in the total taxa richness, number of genera, number of Odonata taxa and the number of Coleoptera taxa ($p < 0.001$) between humic lakes with natural and altered water level. We selected Coleoptera taxa richness to include into the multimetric index. Additionally, Šidagytė et al. [69] demonstrated strong relationship between Coleoptera taxa richness and different pressure descriptors, both eutrophication and hydromorphological alterations.

### 4.1.3. Composition/Abundance Metrics

Composition/abundance metrics are widely used in lake assessment systems, mostly expressed as proportion (relative abundance) of specific taxa. Thus, Lithuanian and Danish assessment systems include % COP (Coleoptera, Odonata, Plecoptera) but Greek littoral assessment system—% of Diptera, and German assessment system—% Odonata. Several studies demonstrated between proportion of different taxa % Gastropoda, % Odonata and anthropogenic pressures [69,73]. We have found that % Coleoptera, % Odonata and % Diptera differ strikingly between natural and altered lakes; however, abundance of total community (expressed as ind/m$^2$) was selected as a core metric due to the strong inter-correlations between the proportion and the diversity metrics [53].

### 4.1.4. Functional Metrics

An alternative to species identity-based methods is the use of functional metrics based on species traits. This approach is recommended by many studies [75–77]. However, only few lake assessment systems include functional metrics, e.g., % abundance of feeding type collectors, is included in Austrian and German lake assessment systems, and % abundance of habitat type lithal in the German alpine lake assessment system. Percentage of feeding type predators is used in Sweden to assess acidification [78]. In our study, several functional metrics showed the difference between natural and altered lakes: % littoral, % profundal and % littoral + profundal ($p < 0.001$). Percentage of organisms with littoral and profundal preference was selected to be included in the core metrics. We assume that this group consists of generalist organisms that might indicate the altered water level.

### 4.2. Assessment of Hydrological Modifications

Traditionally, lake assessment has focused on eutrophication using primary producers—phytoplankton, macrophytes and, recently phytobenthos communities [79]. However, hydrological alterations, e.g., regulation of lake water level for power production and flood control is among the major anthropogenic disturbances in boreal aquatic ecosystems [80]. Several studies have demonstrated strong effects of water level fluctuations on lake biota, mainly benthic invertebrates, fish fauna and macrophytes [39,80,81]. For instance, Aroviita and Hämäläinen [74] showed a marked decrease in species richness of benthic macroinvertebrate with increasing intensity of water-level regulation, especially for Ephemeroptera, Trichoptera, Coleoptera or Megaloptera. Similarly, changes in species composition were reported in regulated lakes of Ireland, i.e., decrease of Crustaceans, increase of Chironomids, Oligochaetes, and invasive amphipods [82] and in regulated lakes in Italy—increase in mobile and/or feeding opportunistic taxa and decrease in sessile and/or herbivorous taxa [83]. Additionally, Brauns et al. [84] described the decrease in Coleoptera, Odonata, Trichoptera and functional groups of piercers, predators, shredders and xylophagous as the potential effects of lake water level fluctuation in lakes of Germany.

Furthermore, the decrease in benthic invertebrate biomass and/or abundance in the littoral area of regulated lakes has been reported by several studies [85–87]. However, other studies did not find any significant effect on numerical abundance of invertebrates [82], probably because water level regulation exerts stronger effect on the biomass of invertebrates than on the numerical abundance affecting larger taxa more [74].

So far, only two lake ecological assessment systems addressed hydromorphological alterations: Slovenian lake assessment system [38] and German assessment system for alpine lakes [36]. However, several lake assessments include multiple pressures including hydromorphological alterations (e.g., [69]). Similar to other studies, we found a marked change in composition and abundance of benthic invertebrates, which further can be used in the development of assessment systems, specifically addressing effect of hydrological alterations in highly humic lakes.

### 4.3. Assessment of Highly Humic Lakes

Humic lakes constitute a considerable portion of lakes in the boreal zone [1]. However, most assessment systems are developed for clear-water lakes characterized by neutral to alkaline pH, low level of DOM, and water transparency depending on the number of phytoplankton. Humic lakes differ substantially from these clear-water systems and therefore might need different assessment methods [13,17,40]. For instance, Benthic Quality Index (BQI)—a widely used metric in the assessment of lake status in Sweden and Finland [87] was deemed inadequate of assessment of the humic lakes as oxygen depletion and dominance by the tolerant species *Chironomus anthracinus* and *C. plumosus* are natural phenomena and not an effect of human impacts [17]. Further, several phytoplankton, macrophyte and fish metrics:

(i)   Classified reference lakes as impacted;
(ii)  Did not differentiate between reference and impacted lakes [17].

This can be explained by the fact that humic lake communities are more tolerant to the environmental fluctuations [88] and are less taxa-rich and diverse comparing to clear-water communities [8]. Further, light limitation due to high level of humic substances plays an important role in these lakes, so several widely used lake assessment parameters as macrophyte colonization depth cannot be used in these lakes [14].

In Latvia, lake assessment system is based on number of taxa, number of EPTCBO taxa, Shannon-Wiener diversity index, ASPT index and acidity index [89]. However, this assessment system was not appropriate for assessment of humic lakes, as also near-natural lakes were classified as impacted according to this system [50]. This problem was encountered also in lakes of Finland [17]. Unsuitability of certain metrics for highly humic lakes can be solved by developing an assessment system targeted to specific human pressure and lake types, as shown by this study. Whether this multimetric index is applicable to humic lakes in other regions needs to be tested in future studies.

**Supplementary Materials:** The following are available online at https://www.mdpi.com/2073-4441/13/2/223/s1, Table S1: The results of the binary logistic regression (BLR). Z and *p* values of the chosen metrics between natural and altered highly humic lakes, Table S2: Correlation coefficients of the selected metrics for the highly humic lake multimetric index, Table S3: Anchor values of the macroinvertebrate metrics for a potential multimetric index for the highly humic lake quality assessment, Table S4: Macroinvertebrate taxa found in the studied highly humic lakes.

**Author Contributions:** Conceptualization, D.O., S.P., I.K.; funding acquisition, D.O.; field works, D.O., A.S., J.J.; sample analysis, D.O., A.S.; data analysis, A.A., D.O., A.S., I.K., J.J.; Writing of original draft, D.O., S.P.; Review and editing, I.K., J.J., A.A., A.S. All authors have read and agreed to the published version of the manuscript.

**Funding:** The data were collected in the frame of ESF supported project PuReST, No. 1DP/1.1.1.2.0/13/APIA/VIAA/044 and a project supported by The Administration of Latvian Environmental Protection Fund, No. 1-08/202/2018.

**Institutional Review Board Statement:** Not applicable.

**Informed Consent Statement:** Not applicable.

**Data Availability Statement:** The data presented in this study are available on request from the corresponding author. The data are not publicly available due to the data acquisition from two different projects and the National Surface Water Monitoring program, all requiring different data availability policy.

**Acknowledgments:** Authors want to express their gratitude to the Latvian Environment, Geology and Meteorology Centre.

**Conflicts of Interest:** The authors declare no conflict of interest.

## References

1. Keskitalo, J.; Eloranta, P. (Eds.) *Limnology of Humic Waters*; Backhuys Publishers: Leiden, The Netherlands, 1999; p. 284.
2. Wetzel, R.G. *Limnology: Lake and River Ecosystems*, 3rd ed.; Academic Press: San Diego, CA, USA, 2001; p. 1006.
3. Salonen, K.; Arvola, L.; Tulonen, T.; Hammar, T.; Metsälä, T.R.; Kankaala, P.; Münster, U. Planktonic food chains of a highly humic lake. I. A mesocosm experiment during the spring primary production maximum. *Hydrobiology* **1992**, *229*, 125–142. [CrossRef]
4. Jones, R.I. The influence of humic substances on lacustrine planktonic food-chains. *Hydrobiology* **1992**, *229*, 73–91. [CrossRef]
5. Solomon, C.T.; Jones, S.E.; Weidel, B.C.; Buffam, I.; Fork, M.L.; Karlsson, J.; Larsen, S.; Lennon, J.T.; Read, J.S.; Sadro, S.; et al. Ecosystem consequences of changing inputs of terrestrial dissolved organic matter to lakes: Current knowledge and future challenges. *Ecosystem* **2015**, *18*, 376–389. [CrossRef]
6. Steinberg, C.E.W.; Kamara, S.; Prokhotskaya, V.Y.; Manusadzianas, L.; Karasyova, T.A.; Timofeyev, M.A.; Jie, Z.; Paul, A.; Meinelt, T.; Farjalla, V.F.; et al. Dissolved humic substances—Ecological driving forces from the individual to the ecosystem level? *Freshw. Biol.* **2006**, *51*, 1189–1210. [CrossRef]
7. Lepistö, L.; Holopainen, A.L.; Vuoristo, H. Type-specific and indicator taxa of phytoplankton as a quality criterion for assessing the ecological status of Finnish boreal lakes. *Limnology* **2004**, *34*, 236–248. [CrossRef]
8. Toivonen, H.; Huttunen, P. Aquatic macrophytes and ecological gradients in 57 small lakes in southern Finland. *Aquat. Bot.* **1995**, *51*, 197–221. [CrossRef]
9. Vesterinen, J.; Devlin, S.P.; Syväranta, J.; Jones, R.I. Influence of littoral periphyton on whole-lake metabolism relates to littoral vegetation in humic lakes. *Ecology* **2017**, *98*, 3074–3085. [CrossRef]
10. Sarvala, J.; Kankaala, P.; Zingel, P.; Arvola, L. Zooplankton. In *Limnology of Humic Waters*; Keskitalo, J., Eloranta, P., Eds.; Backhuys Publishers: Leiden, The Netherlands, 1999; pp. 173–191.
11. Rask, M.; Viljanen, M.; Sarval, J. Humic lakes as fish habitats. In *Limnology of Humic Waters*; Keskitalo, J., Eloranta, P., Eds.; Backhuys Publishers: Leiden, The Netherlands, 1999; pp. 209–224.
12. Ptacnik, R.; Solimini, A.G.; Brettum, P. Performance of a new phytoplankton composition metric along a eutrophication gradient in Nordic lakes. *Hydrobiology* **2009**, *633*, 75–82. [CrossRef]
13. Rask, M.; Vuori, K.-M.; Hämäläinen, H.; Järvinen, M.; Hellsten, S.; Mykrä, H.; Arvola, L.; Ruuhijärvi, J.; Jyväsjärvi, J.; Kolari, I.; et al. Ecological classification of large lakes in Finland: Comparison of classification approaches using multiple quality elements. *Hydrobiology* **2011**, *660*, 37–47. [CrossRef]
14. Søndergaard, M.; Phillips, G.; Hellsten, S.; Kolada, A.; Ecke, F.; Mäemets, H.; Mjelde, M.; Azzella, M.M.; Oggioni, A. Maximum growing depth of submerged macrophytes in European lakes. *Hydrobiology* **2013**, *704*, 165–177. [CrossRef]
15. Lepistö, L.; Holopainen, A.L.; Vuoristo, H.; Rekolainen, S. Phytoplankton assemblages as a criterion in the ecological classification of lakes in Finland. *Boreal Environ. Res.* **2006**, *11*, 35–44.
16. Birk, S.; Ecke, F. The potential of remote sensing in ecological status assessment of coloured lakes using aquatic plants. *Ecol. Indic.* **2014**, *46*, 398–406. [CrossRef]
17. Alahuhta, J.; Vuori, K.M.; Hellsten, S.; Järvinen, M.; Olin, M.; Rask, M.; Palomäki, A. Defining the ecological status of small forest lakes using multiple biological quality elements and palaeolimnological analysis. *Fund. Appl. Limnol.* **2009**, *175*, 203–216. [CrossRef]
18. Baars, J.R.; Murray, D.A.; Hannigan, E.; Kelly-Quinn, M. Macroinvertebrate assemblages of small upland peatland lakes in Ireland. *Biol. Environ.* **2014**, *114*, 233–248. [CrossRef]
19. Armstrong, A.; Holden, J.; Kay, P.; Francis, B.; Foulger, M.; Gledhill, S.; McDonald, A.T.; Walker, A. The impact of peatland drain-blocking on dissolved organic carbon loss and discolouration of water; results from a national survey. *J. Hydrol.* **2010**, *381*, 112–120. [CrossRef]
20. Beadle, J.M.; Brown, L.E.; Holden, J. Biodiversity and ecosystem functioning in natural bog pools and those created by rewetting schemes. *Wiley Interdiscip. Rev. Water* **2015**, *2*, 65–84. [CrossRef]
21. Lepistö, L.; Saura, M. Effects of forest fertilization on phytoplankton in a boreal brown-water lake. *Boreal Environ. Res.* **1998**, *3*, 33–43.
22. Drinan, T.J.; Graham, C.T.; O'Halloran, J.; Harrison, S.S.C. The impact of conifer plantation forestry on the Chydoridae (Cladocera) communities of peatland lakes. *Hydrobiology* **2013**, *700*, 203–219. [CrossRef]
23. Deininger, A.; Jonsson, A.; Karlsson, J.; Bergström, A.K. Pelagic food webs of humic lakes show low short-term response to forest harvesting. *Ecol. Appl.* **2018**, *29*, 1–12. [CrossRef]
24. Drinan, T.J.; Foster, G.N.; Nelson, B.H.; O'Halloran, J.; Harrison, S.S.C. Macroinvertebrate assemblages of peatland lakes: Assessment of conservation value with respect to anthropogenic land-cover change. *Biol. Conserv.* **2013**, *158*, 175–187. [CrossRef]

25. Graham, C.T.; Drinan, T.J.; Harrison, S.S.; O'Halloran, J. Relationship between plantation forest and brown trout growth, energetics and population structure in peatland lakes in western Ireland. *For. Ecol. Manag.* **2014**, *321*, 71–80. [CrossRef]
26. European Council Directive. European Commission Directive 2000/60/EC of the European Parliament and of the Council of 23 October 2000 Establishing a Framework for Community Action in the Field of Water Policy. *OJEC* **2000**, *327*, 1–72.
27. Poikane, S.; Salas Herrero, F.; Kelly, M.G.; Borja, A.; Birk, S.; van de Bund, W. European aquatic ecological assessment methods: A critical review of their sensitivity to key pressures. *Sci. Total Environ.* **2020**, *740*, 140075. [CrossRef] [PubMed]
28. Poikane, S.; Zampoukas, N.; Borja, A.; Davies, S.P.; van de Bund, W.; Birk, S. Intercalibration of aquatic ecological assessment methods in the European Union: Lessons learned and way forward. *Environ. Sci. Policy* **2014**, *44*, 237–246. [CrossRef]
29. Poikane, S.; Birk, S.; Böhmer, J.; Carvalho, L.; de Hoyos, C.; Gassner, H.; Hellsten, S.; Kelly, M.; Solheim, A.L.; Olin, M.; et al. A hitchhiker's guide to European lake ecological assessment and intercalibration. *Ecol. Indic.* **2015**, *52*, 533–544. [CrossRef]
30. Carvalho, L.; Poikane, S.; Solheim, A.L.; Phillips, G.; Borics, G.; Catalan, J.; De Hoyos, C.; Drakare, S.; Dudley, B.J.; Järvinen, M.; et al. Strength and uncertainty of phytoplankton metrics for assessing eutrophication impacts in lakes. *Hydrobiology* **2013**, *704*, 127–140. [CrossRef]
31. Phillips, G.; Lyche-Solheim, A.; Skjelbred, B.; Mischke, U.; Drakare, S.; Free, G.; Järvinen, M.; de Hoyos, C.; Morabito, G.; Poikane, S.; et al. A phytoplankton trophic index to assess the status of lakes for the Water Framework Directive. *Hydrobiology* **2013**, *704*, 75–95. [CrossRef]
32. Penning, W.E.; Dudley, B.; Mjelde, M.; Hellsten, S.; Hanganu, J.; Kolada, A.; van den Berg, M.; Poikane, S.; Phillips, G.; Willby, N.; et al. Using aquatic macrophyte community indices to define the ecological status of European lakes. *Aquat. Ecol.* **2008**, *42*, 253–264. [CrossRef]
33. Poikane, S.; Portielje, R.; Denys, L.; Elferts, D.; Kelly, M.; Kolada, A.; Mäemets, H.; Phillips, G.; Søndergaard, M.; Willby, N.; et al. Macrophyte assessment in European lakes: Diverse approaches but convergent views of 'good' ecological status. *Ecol. Indic.* **2018**, *94*, 185–197. [CrossRef]
34. Kelly, M.; Urbanic, G.; Acs, E.; Bennion, H.; Bertrin, V.; Burgess, A.; Denys, L.; Gottschalk, S.; Kahlert, M.; Karjalainen, S.M.; et al. Comparing aspirations: Intercalibration of ecological status concepts across European lakes for littoral diatoms. *Hydrobiology* **2014**, *734*, 125–141. [CrossRef]
35. Lyche-Solheim, A.; Feld, C.K.; Birk, S.; Phillips, G.; Carvalho, L.; Morabito, G.; Mischke, U.; Willby, N.; Søndergaard, M.; Hellsten, S.; et al. Ecological status assessment of European lakes: A comparison of metrics for phytoplankton, macrophytes, benthic invertebrates and fish. *Hydrobiology* **2013**, *704*, 57–74. [CrossRef]
36. Poikane, S.; Johnson, R.K.; Sandin, L.; Schartau, A.K.; Solimini, A.G.; Urbanič, G.; Arbačiauskas, K.; Aroviita, J.; Gabriels, W.; Miler, O.; et al. Benthic macroinvertebrates in lake ecological assessment: A review of methods, intercalibration and practical recommendations. *Sci. Total Environ.* **2016**, *543*, 123–134. [CrossRef] [PubMed]
37. EEA. *European Waters—Assessment of Status and Pressures*; EEA Report; European Environment Agency: København, Denmark, 2018; pp. 1–90. [CrossRef]
38. Urbanič, G. A Littoral Fauna Index for assessing the impact of lakeshore alterations in Alpine lakes. *Ecohydrology* **2014**, *7*, 703–716. [CrossRef]
39. Mjelde, M.; Hellsten, S.; Ecke, F. A water level drawdown index for aquatic macrophytes in Nordic lakes. *Hydrobiology* **2013**, *704*, 141–151. [CrossRef]
40. Olin, M.; Rask, M.; Ruuhijärvi, J.; Tammi, J. Development and evaluation of the Finnish fish-based lake classification method. *Hydrobiology* **2013**, *713*, 149–166. [CrossRef]
41. Van Duinen, G.A.; Brock, A.M.; Kuper, J.T.; Leuven, R.S.E.W.; Peeters, T.M.J.; Roelofs, J.G.M.; van der Velde, G.; Verberk, W.C.E.P.; Esselink, H. Do restoration measures rehabilitate fauna diversity in raised bogs? A comparative study on aquatic macroinvertebrates. *Wetl. Ecol. Manag.* **2003**, *11*, 447–459. [CrossRef]
42. Hannigan, E.; Kelly-Quinn, M. Composition and structure of macroinvertebrate communities in contrasting open-water habitats in Irish peatlands: Implications for biodiversity conservation. *Hydrobiology* **2012**, *692*, 19–28. [CrossRef]
43. Walker, I.R.; Fernando, C.H.; Paterson, C.G. Associations of Chironomidae (Diptera) of shallow, acid, humic lakes and bog pools in Atlantic Canada, and a comparison with an earlier paleoecological investigation. *Hydrobiology* **1985**, *120*, 11–22. [CrossRef]
44. Pakulnicka, J.; Zawal, A. Model of disharmonic succession of dystrophic lakes based on aquatic beetle fauna (Coleoptera). *Mar. Freshw. Res.* **2019**, *70*, 195–211. [CrossRef]
45. Pakalne, M.; Kalnina, L. Mire ecosystems in Latvia. *Stapfia* **2005**, *85*, 147–174.
46. Montanarella, L.; Jones, R.J.A.; Hiederer, R. The distribution of peatland in Europe. *Mires Peat* **2006**, *1*, 1–10.
47. European Council Directive. Council Directive 92/43/EEC on the conservation of natural habitats and of wild fauna and flora. *OJEC* **1992**, *206*, 7–42.
48. Pakalne, M.; Kalnina, L. Mires in Latvia. *Suo* **2000**, *51*, 213–226.
49. Klavins, M.; Kokorite, I.; Springe, G.; Skuja, A.; Parele, E.; Rodinov, V.; Druvietis, I.; Strāķe, S.; Urtans, A. Water quality in cutaway peatland lakes in Seda mire, Latvia. *Ecohydrol. Hydrobiol.* **2010**, *10*, 61–70. [CrossRef]
50. Latvian Environment, Geology and Meteorology Centre. River Basin Management Plans 2016–2021. Available online: https://videscentrs.lvgmc.lv/lapas/udens-apsaimniekosana-un-pludu-parvaldiba (accessed on 7 December 2020).
51. American Public Health Association; American Water Works Association; Water Environment Federation. *Standard Methods for the Examination of Water and Wastewater*, 21th ed.; American Public Health Association: Washington, DC, USA, 2005; p. 1496.

52. AQEM European Stream Assessment Program. English Manual, Version 2.3. Available online: http://www.aqem.de/mains/products.php (accessed on 7 December 2020).

53. Hering, D.; Feld, C.K.; Moog, O.; Ofenbock, T. Cook book for the development of a Multimetric Index for biological condition of aquatic ecosystems: Experiences from the European AQEM and STAR projects and related initiatives. *Hydrobiology* **2006**, *566*, 311–324. [CrossRef]

54. Sokal, R.R.; Rohlf, F.J. *Biometry: The Principles and Practice of Statistics in Biological Research*, 3rd ed.; W.H. Freeman and Co: New York, NY, USA, 1995; p. 880.

55. R Core Team. R: A Language and Environment for Statistical Computing. 2020. Available online: http://www.r-project.org/index.html (accessed on 3 November 2020).

56. Wickham, H.; Averick, M.; Bryan, J.; Chang, W.; McGowan, L.; François, R.; Grolemund, G.; Hayes, A.; Henry, L.; Hester, J.; et al. Welcome to the Tidyverse. *J. Open Source Softw.* **2019**, *4*, 1686. [CrossRef]

57. Convention on the Conservation of European Wildlife and Natural Habitats (Bern, Switzerland). 1979. Available online: https://rm.coe.int/1680078aff (accessed on 12 January 2021).

58. Desrochers, A.; van Duinen, G.J. Peatland fauna. In *Boreal Peatland Ecosystems, Ecological Studies 188*; Wieder, R.K., Vitt, D.H., Eds.; Springer: Berlin/Heidelberg, Germany, 2006; pp. 67–100.

59. Ewald, M.L.; Feminella, J.W.; Lenertz, K.K.; Henry, R.P. Acute physiological responses of the freshwater snail Elimia flava (Mollusca: Pleuroceridae) to environmental pH and calcium. *Comp. Biochem. Physiol. Part C: Toxicol. Pharmacol.* **2009**, *150*, 237–245. [CrossRef]

60. Mäkelä, T.P.; Oikari, A.O.J. The effects of low water pH on the ionic balance in freshwater mussel Anodonta anatina L. *Ann. Zool. Fennici.* **1992**, *29*, 169–175.

61. Schartau, A.K.; Moe, S.J.; Sandin, L.; McFarland, B.; Raddum, G.G. Macroinvertebrate indicators of lake acidification: Analysis of monitoring data from UK, Norway and Sweden. *Aquat. Ecol.* **2008**, *42*, 293–305. [CrossRef]

62. Soldán, T.; Bojková, J.; Vrba, J.; Bitušík, P.; Chvojka, P.; Papáček, M.; Peltanová, J.; Sychra, J.; Tátosová, J. Aquatic insects of the Bohemian Forest glacial lakes: Diversity, long-term changes, and influence of acidification. *Silva Gabreta* **2012**, *18*, 123–283.

63. Karr, J.R.; Chu, E.W. *Biological Monitoring and Assessment: Using Multimetric Indexes Effectively*; EPA 235-R97-001; US Environmental Protection Agency: Washington, DC, USA, 1997; p. 149.

64. Armitage, P.D.; Moss, D.; Wright, J.F.; Furse, M.T. The performance of a new biological water quality score system based on macroinvertebrates over a wide range of unpolluted running-water sites. *Water Res.* **1983**, *17*, 333–347. [CrossRef]

65. Davy-Bowker, J.; Murphy, J.F.; Rutt, G.P.; Steel, J.E.C.; Furse, M.T. The development and testing of a macroinvertebrate biotic index for detecting the impact of acidity on streams. *Arch. Hydrobiol.* **2005**, *163*, 383–403. [CrossRef]

66. Wiederholm, T. Use of benthos in lake monitoring. *J. Water Pollut. Con. Fed.* **1980**, *52*, 537–547.

67. Miler, O.; Porst, G.; McGoff, E.; Pilotto, F.; Donohue, L.; Jurca, T.; Solimini, A.G.; Sandin, L.; Irvin, K.; Aroviita, J.; et al. Morphological alterations of lake shores in Europe: A multimetric ecological assessment approach using benthic macroinvertebrates. *Ecol. Indic.* **2013**, *34*, 398–410. [CrossRef]

68. Hawkes, H.A. Origin and development of the biological monitoring working party score system. *Water Res.* **1998**, *32*, 964–968. [CrossRef]

69. Šidagytė, E.; Višinskienė, G.; Arbačiauskas, K. Macroinvertebrate metrics and their integration for assessing the ecological status and biocontamination of Lithuanian lakes. *Limnologica* **2013**, *43*, 308–318. [CrossRef]

70. Mavromati, E.; Kemitzoglou, D.; Tsiaoussi, V.; Lazaridou, M. *Report on the Development of the National Method for the Assessment of Ecological Status of Natural Lakes in Greece, with the Use of Littoral Benthic Invertebrates*; Greek Biotope/Wetland Centre and Special Secretariat for the Natural Environment and Waters, Ministry of Environment and Energy: Thermi, Greece, 2020.

71. Wiberg-Larsen, P.; Rasmussen, J.J. Revised Danish Macroinvertebrate Index for Lakes—A Method to Assess Ecological Quality. Aarhus University, DCE—Danish Centre for Environment and Energy. 2020. Available online: http://dce2.au.dk/pub/SR373.pdf (accessed on 17 January 2020).

72. Böhmer, J.; Chiriac, G.; Varbiro, G.; Wolfram, G.; Poikane, S. *Intercalibrating the National Classifications of Ecological Status for Eastern Continental Lakes: Biological Quality Element: Benthic Invertebrates*; EUR 29342 EN; Publications Office of the European Union: Luxembourg, 2018; p. 54. [CrossRef]

73. Urbanič, G.; Petkovska, V.; Pavlin, M. The relationship between littoral benthic invertebrates and lakeshore modification pressure in two alpine lakes. *Fundam. Appl. Limnol.* **2012**, *180*, 157–173. [CrossRef]

74. Aroviita, J.; Hämäläinen, H. The impact of water-level regulation on littoral macroinvertebrate assemblages in boreal lakes. *Hydrobiology* **2008**, *613*, 45–56. [CrossRef]

75. Friberg, N.; Bonada, N.; Bradley, D.C.; Dunbar, M.J.; Edwards, F.K.; Grey, J.; Hayes, R.B.; Hildrew, A.G.; Lamouroux, N.; Trimmer, M.; et al. Biomonitoring of human impacts in freshwater ecosystems. The good, the bad and the ugly. *Adv. Ecol. Res.* **2011**, *44*, 1–68. [CrossRef]

76. Reyjol, Y.; Argillier, C.; Bonne, W.; Borja, A.; Buijse, A.D.; Cardoso, A.C.; Daufresne, M.; Kernan, M.; Ferreira, M.T.; Poikane, S.; et al. Assessing the ecological status in the context of the European Water Framework Directive: Where do we go now? *Sci. Total Environ.* **2014**, *497*, 332–344. [CrossRef] [PubMed]

77. Brucet, S.; Poikane, S.; Lyche-Solheim, A.; Birk, S. Biological assessment methods for European lakes: Ecological rationale and human impacts. *Freshw. Biol.* **2013**, *58*, 1106–1115. [CrossRef]

78. Johnson, R.K.; Goedkoop, W. *Bedömningsgrunder för Bottenfauna i Sjöar Och Vattendrag: Användarmanual och Bakgrundsdokument*; Sveriges lantbruksuniversitet: Uppsala, Sweden, 2007; p. 84.

79. Kelly, M.G.; Birk, S.; Willby, N.J.; Denys, L.; Drakare, S.; Kahlert, M.; Karjalainen, S.M.; Marchetto, A.; Pitt, J.A.; Urbanič, G.; et al. Redundancy in the ecological assessment of lakes: Are phytoplankton, macrophytes and phytobenthos all necessary? *Sci. Total Environ.* **2016**, *568*, 594–602. [CrossRef] [PubMed]

80. Sutela, T.; Vehanen, T.; Rask, M. Assessment of the ecological status of regulated lakes: Stressor-specific metrics from littoral fish assemblages. *Hydrobiology* **2011**, *675*, 55–64. [CrossRef]

81. Sutela, T.; Aroviita, J.; Keto, A. Assessing ecological status of regulated lakes with littoral macrophyte, macroinvertebrate and fish assemblages. *Ecol. Indic.* **2013**, *24*, 185–192. [CrossRef]

82. Evtimova, V.V.; Donohue, I. Water-level fluctuations regulate the structure and functioning of natural lakes. *Freshw. Biol.* **2016**, *61*, 251–264. [CrossRef]

83. Mastrantuono, L.; Pilotto, F.; Rossopinti, A.; Bazzanti, M.; Solimini, G. Response of littoral macroinvertebrates to morphological disturbances in Mediterranean lakes: The case of Lake Piediluco (Central Italy). *Fund. Appl. Limnol.* **2015**, *186*, 297–310. [CrossRef]

84. Brauns, M.; Garcia, X.F.; Pusch, M. Potential effects of water level fluctuations on littoral invertebrates in lowland lakes. *Hydrobiology* **2008**, *613*, 5–12. [CrossRef]

85. Palomäki, R. Response by macrozoobenthos biomass to water level regulation in some Finnish lake littoral zones. *Hydrobiology* **1994**, *286*, 17–26. [CrossRef]

86. Tikkanen, P.; Kantola, L.; Niva, T.; Hellsten, S.; Alasaarela, E. *Ecological Aspects of Lake Regulation in Northern Finland. Part 3. Macrozoobenthos and Feeding of Fish*; VTT Research Notes 987; Technical Research Centre of Finland: Espoo, Finland, 1989; (in Finnish with English).

87. Jyväsjärvi, J.; Nyblom, J.; Hämäläinen, H. Palaeolimnological validation of estimated reference values for a lake profundal macroinvertebrate metric (Benthic Quality Index). *J. Paleolimnol.* **2010**, *44*, 253–264. [CrossRef]

88. Tammi, J.; Lappalainen, A.; Mannio, J.; Rask, M.; Vuorenmaa, J. Effects of eutrophication on fish and fisheries in Finnish lakes: A survey based on random sampling. *Fish. Manag. Ecol.* **1999**, *6*, 173–186. [CrossRef]

89. Skuja, A.; Ozoliņš, D. *Fitting New Method—Latvian Lake Macroinvertebrate Multimetric Index (LLMMI) to Results of Central—Baltic Geographical Intercalibration Group (CB—GIG) Lake Benthic Macroinvertebrate Intercalibration*; Report; Institute of Biology, University of Latvia: Salaspils, Latvia, 2016; p. 12.

 *water*

*Article*

# Alien Crayfish Species in the Deep Subalpine Lake Maggiore (NW-Italy), with a Focus on the Biometry and Habitat Preferences of the Spiny-Cheek Crayfish

**Laura Garzoli**, **Stefano Mammola**, **Marzia Ciampittiello** and **Angela Boggero** *

CNR-Water Research Institute (CNR-IRSA), Corso Tonolli 50, 28922 Verbania, Italy;
laura.garzoli01@gmail.com (L.G.); stefanomammola@gmail.com (S.M.); marzia.ciampittiello@irsa.cnr.it (M.C.)
* Correspondence: angela.boggero@irsa.cnr.it; Tel.: +39-03-0323-518300

Received: 29 March 2020; Accepted: 12 May 2020; Published: 14 May 2020

**Abstract:** Invasive alien species are a major threat to biodiversity. Thus, it is fundamental to implement control strategies at the early stages of invasions. In the framework of the Italian-Swiss Alien Invasive Species in Lake Maggiore cooperative programme, we performed an extensive study on the occurrence and ecology of alien crayfish, one of the most significant invaders of freshwater habitats. From April 2017 to July 2018, we inspected seventy-five sites along the coastline to verify crayfish occurrence. We recorded, for the first time, the signal crayfish *Pacifastacus leniusculus*. Additionally, we found few individuals and remains of the red swamp crayfish *Procambarus clarkii*, and confirmed the presence of a consistent population of the spinycheek crayfish *Orconectes limosus*. Given the high number of *O. limosus'* individuals found, it was possible to perform in-depth biometric and ecological analyses for this abundant species only. We observed no significant differences of biometric measures between males and females of *O. limosus*. We explore its habitat preferences with a generalized linear model, detecting a significant relationship between mean annual temperatures and the presence of shelters of this species. These results, together, have direct implications for planning rapid management response actions on alien crayfish in large and deep lakes.

**Keywords:** lakes; invasion biology; non-indigenous species; *Procambarus Clarkii*; *Pacifastacus Leniusculus*; *Orconectes Limosus*

## 1. Introduction

Invasive Alien Species (IAS) are regarded as one of the major drivers of the global biodiversity decline [1–3]. They impact native species and ecosystems, but also the human society, by threatening Nature's Contribution to People (NCP) and causing economic and cultural damages [4]. In Europe, crayfish represent the most frequent freshwater invaders, responsible for local extinctions of native species, damage to freshwater resources, as well as to productive activities [5,6]. Their introduction is a serious threat to these key habitats [7], which are usually already compromised by several anthropogenic stressors including chemical pollution, climate change, and water derivations and withdrawals [8,9]. The spread of IAS is a further impeding factor that limits the effectiveness of protection, vanishing the areas of concern management, and the restoration strategies [10]. In Europe, there are at least eight species of alien crayfish, introduced intentionally or accidentally during the nineteenth and twentieth centuries [11]. Five of these species are considered of Union Concern [12]: *Orconectes limosus* (Rafinesque, 1817), *O. virilis* (Hagen, 1870), *Pacifastacus leniusculus* (Dana, 1852), *Procambarus clarkii* (Girard, 1852), and *Procambarus fallax* (Hagen, 1870) f. *virginalis*.

During the last 50 years, Italian lakes and rivers have been heavily affected by anthropogenic pressures, including the arrival of neobiota that have altered the native communities [13,14], and have caused major impacts on the environment, as well as on socio-economic activities. Due to the complexity of their food webs structures and their socio-economic relevance, deep lakes deserve to be studied using deeper and articulated approaches than the standard routine (Water Framework Directive 2000/60/EC), in order to obtain a more nuanced understanding of their short- to long-term evolution.

Lake Maggiore is a large and deep lake of fluvio-glacial origin located on the southern side of the Western Alps, straddling the Italian-Swiss border. Over the past century, the catchment area of the lake has been interested by intense urbanization and industrial activities, which contributed to chemical and organic pollution of lake waters. In the past, eutrophication was one of the primary impacts to the lake ecosystem [15], followed by the rise in temperature of 1.4 °C in the upper 30 m depth (historical period 1965–2010). The synergistic effect of these two factors resulted in the extension of the epilimnion at lower depth and an increase in harmful cyanobacteria blooms frequency [16]. All these changes also favoured IAS arrival and acclimatization in the lake [17–20]. Two species of alien crayfish are reported to inhabit the lake: The red swamp crayfish *Procambarus clarkii* [21] and the spiny-cheek crayfish *Orconectes limosus* [22]. Since the 1970s, the lake has been included in the monitoring program of the International Commission for the Protection of Italian-Swiss Waters (CIPAIS), which provides periodic monitoring of biological (phyto- and zooplankton, fish) and of physical-chemical parameters along the water column. Using this lake as a model system, in 2017 the project "Alien Invasive Species in Lake Maggiore" (SPAM) began, aiming at documenting the occurrence, the spatial distribution, and the abundance of alien macrophytes, crayfish and bivalves along the lake shores.

Within the SPAM framework, we conducted an extensive monitoring of the coastline of Lake Maggiore, aiming to identify the occurrence of alien crayfish. Once we identified the resident crayfish within the lake, we conducted an in-depth analysis focused on the most abundant species, to shed light on its natural history and autoecology. Specifically, we acquired biometric data on *O. limosus* in order to describe population size and variation in body traits, across the different life stages, and we studied the most important drivers of its spatial distribution. Building upon this evidence, our over-arching goal is to provide indications for the long-term management strategies to control alien crayfish in Lake Maggiore.

## 2. Materials and Methods

### 2.1. Study Site

Located in NW-Italy, Lake Maggiore is the second Italian lake by extension (area of 212 km$^2$, maximum depth 370 m, volume of 37.5 km$^3$) [17]. Its surface area partly belongs to the Italian territory (80%), and partly to Switzerland (20%), while its catchment (6599 km$^2$) is roughly equally divided between the two Countries. Due to the presence of the Alps, the local climate is characterised by high mean annual precipitation (1658 mm; reference period 1981–2018), with increasingly frequent extreme events primarily concentrated in the last 20 years [23].

### 2.2. Sampling Procedures

We monitored the whole coastline of Lake Maggiore for the occurrence of alien crayfish. In 2017, we carried out preliminary inspections to record the environmental features of the shorelines and to develop a standardized sampling protocol. We set the minimum distance between two different recording stations at 1.5–2 km along the coastline, to cover the entire perimeter of the lake (Figure 1), for a total of 75 sampling stations. We evaluated the following environmental features: type of substrate, dominant vegetation of the banks and of the shores, and the presence of native and alien crayfish (traces, burrows, remains, or live individuals). We classified the type of substrate basing on their granulometry as sand/silt (particle size < 1 mm), pebbles (particle size range 2-256 mm), and boulders (particle size > 256 mm) [24]. We assessed the occurrence of alien crayfish by visual encounter surveys [25] in the sites visited for preliminary inspections. We standardize research efforts by setting visual assessment time at 20 min/site [25]. In parallel, local fishermen were interviewed to gain anecdotal data on the distribution of the species.

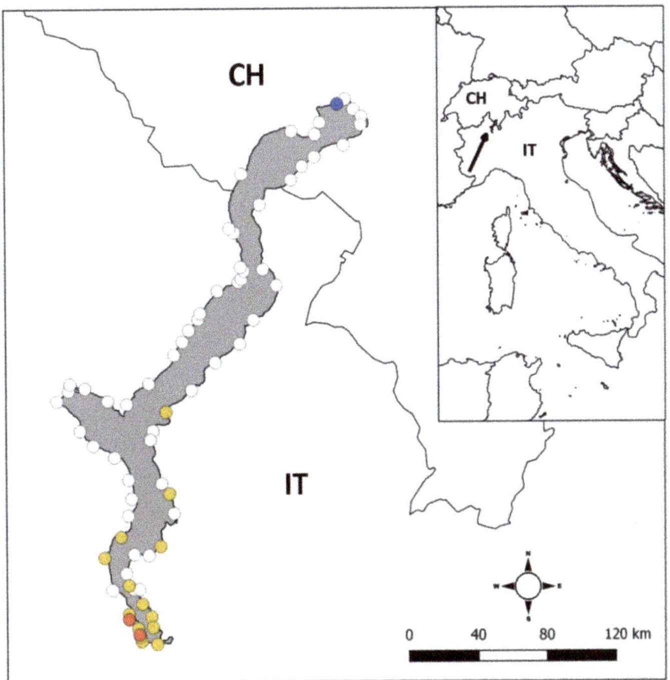

**Figure 1.** Sampling sites (dots) across Lake Maggiore (IT: Italy, CH: Switzerland). Red dots: sites of occurrence of *Procambarus clarkii*; Yellow dots: *Orconectes limosus*; Blue dot: *Pacifastacus leniusculus*; White dots: no alien crayfish detected.

In a second step, we selected eight sites on both eastern and western lake sides, to confirm the previous findings and for a deeper investigation through trapping, by drawing an imaginary line of representative sites along the north-south axis of the lake. We carried out sampling procedures tailoring the protocol described in Tricarico and coauthors [26] to a deep and large lake, placing 25 wire mesh double entrances cylindrical traps (30 × 60 cm) for each site, at a regular distance of ca. 10 m to one another. We used approx. 20 g cat food/trap as a bite to attract crayfish. We checked and removed traps after 24 h to: (i) Avoid wounds and cannibalism phenomena between individuals due to an excessive permanence; (ii) allow the release of by-catches; (iii) avoid the removal of traps by unauthorized personnel.

### 2.3. Biometric Analysis

We identified crayfish taxonomically according to Mazzoni and coauthors [27] and Souty-Grosset and coauthors [28]. For each captured specimen, we recorded: (i) sex; (ii) total body length (Ltot, from the tip of the rostrum to the tip of the telson); (iii) length of the cephalothorax (Lcft, from the tip of the rostrum to the end of the carapace); (iv) total individual weight (Wtot); (v) presence and number of eggs in ovigerous females. Crayfish belonging to the species *O. limosus* were divided in age/size classes [29] (Table 1):

**Table 1.** Age-size classes of *Orconectes limosus* according to Pieplow [29].

| Age | Body Length (mm) |
|-----|------------------|
| 0+ | Up to 40–65 |
| 1+ | 65–80 |
| 2+ | 80–95 |
| 3+ | 95–110 |

*2.4. Statistical Analyses and Models*

We performed all analyses in R software, version 2.13.2 [30]. *Orconectes limosus* was the only abundant species recorded along the lake shores, and thus, all in-depth analyses refer to this species. We tested the differences in biometry using a Student's t-test, taking into consideration the effect of size classes, and comparing present-day biometric data with historical data available in Bazzoni [22]. We graphically compared data using density plots in ggplot2 [31].

To understand environmental preferences of the species, we obtained present-day climatic data on average annual temperatures, maximum annual temperatures, and minimum annual temperatures from WorldClim 2 [32], all variables at a resolution of 30 arc-sec (ca. 1 km at the equator). We derived precipitation data from daily surveys of Lake Maggiore catchment areas used for the preparation of reports for CIPAIS (http://www.cipais.org, accessed on 13 May 2020) and related to the seasonality preceding the sampling period (December 2016–December 2017). We then stacked these environmental variables in a single raster and extracted the climatic conditions for each sampling location.

We defined the habitat preferences of *O. limosus* using regression-type analysis (generalized linear model; GLM) [33]. In contrast to univariate analyses, the use of GLM allowed us to account for the combined effect of explanatory variables as well as potential interactions [33–35]. Given the low abundance of individuals detected in each sampling site (ranging 1–3 individuals), we expressed counts as presence/absence (i.e., Bernoulli distribution 0–1). Thus, we modeled the probability of occurrence rather than abundance values.

Prior to model fitting, we explored the dataset following the protocol for data exploration by Zuur and coauthors [34]. We checked for the presence of outliers using Cleveland's dotplots and we investigated multi-collinearity among continuous covariates using pairwise Pearson correlation tests (r), setting the threshold for collinearity at |r| > 0.7. We inferred the associations between categorical and continuous covariates graphically, with boxplots. Finally, we used coplots to explore possible interaction among covariates.

We developed Bernoulli GLMs in R [30] using a complementary log-log link function (clog-log) as recommended in Zuur and coauthors [35] for datasets with unbalanced presence/absences data (in our case, ~70% observations were absences). Once we fitted the initial model, including all covariates and interactions of interest, we applied model selection [36,37]. We carried out model reduction (backward elimination) on the full model by sequentially deleting terms according to corrected Akaike criterion for finite sample size (AICc) values [38]. We reiterated the reduction process until a minimum adequate model remained, namely the best model supported by observations that avoided overfitting [39].

## 3. Results

*3.1. Occurrence of Alien Crayfish*

We confirmed the occurrence of three alien crayfish in Lake Maggiore. The map of distribution is presented in Figure 1. We recorded the occurrence of the signal crayfish *Pacifastacus leniusculus* for the first time in Lake Maggiore (Canton Tessin, Switzerland) based on three specimens collected in Tenero (Mappo and Rivapiana Minusio) by a professional fisherman [27]. The species was not found in the Italian side of the lake.

We detected a live individual of the red swamp crayfish *Procambarus clarkii* (Palude Bruschera Nature Reserve SCI IT2010015-Lombardy, Italy), while we found only remains in two Dormelletto sites (Piedmont, Italy).

As anticipated in the methods, *O. limosus* was the most abundant invasive crayfish (Figure 1). We found alive specimens in 13 sites along the central-southern Italian shoreline (17.3% of investigated sites), while in 7 additional sites we confirmed the species presence through remains (9.3% of sites).

### 3.2. Biometric Analyses and Population Size Structure of O. Limosus

We collected a total of 238 specimens of *O. limosus*, 72% males and 28% females. A total of 157 crayfish were collected through trapping in 3 out of 8 selected sites (66% of total individuals). In the other five sites there were no stable populations. A summary of biometrics is available in Table 2.

**Table 2.** Body size measures. Distribution range (minimum, maximum, mean, and standard deviation values) for the main biometric features for *Orconectes limosus*. Wtot: $n = 227$, Ltot and Lcft: $n = 238$.

| Sex | Total Body Lenght | | | Total Weight | | | Cephalo-Thorax Length | | |
|---|---|---|---|---|---|---|---|---|---|
| | L min | L max | L mean | W min | W max | W mean | CL min | CL max | CL mean |
| M | 3.94 | 10.57 | 6.66 ± 1.19 | 1.75 | 27.20 | 10.07 ± 5.37 | 1.3 | 4.54 | 3.15 ± 0.73 |
| F | 3.04 | 9.88 | 6.86 ± 1.71 | 0.80 | 30.43 | 11.30 ± 7.47 | 1.1 | 4.89 | 3.24 ± 0.98 |
| Total M+F | 3.04 | 10.57 | 6.71 ± 1.35 | 0.80 | 30.43 | 10.41 ± 6.03 | 1.1 | 4.89 | 3.18 ± 0.80 |

We observed no significant differences between males and females for either cephalothorax length ($t = -0.72$, $p > 0.05$), total length ($t = -1.01$, $p > 0.05$), or weight ($t = -1.39$, $p > 0.05$), even within size classes (Figure 2).

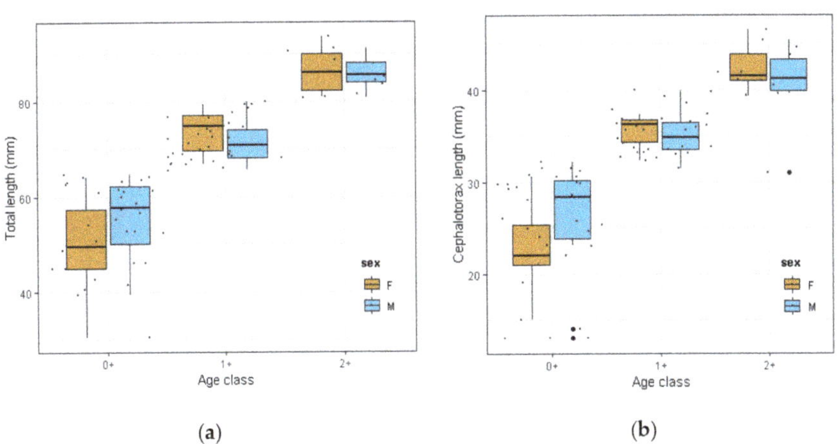


**Figure 2.** Boxplots showing variations in total body length (**a**) and cephalotorax length; (**b**) among age classes of *Orconectes limosus* in Lake Maggiore. Grey dots are observed values used to calculate each boxplot. A random noise (jitter) is applied to aid visualization of otherwise superimposed dots.

The comparisons of data acquired in this study with the ones collected in 2006 by Bazzoni [22], highlighted a significant difference both in total body length ($t = 14.66$, $p < 0.001$) and total weight ($t = 13.36$, $p < 0.001$), and, in the former case, the observed difference is ascribable also to sex ($t = -2.19$, $p = 0.03$) (Figure 3).

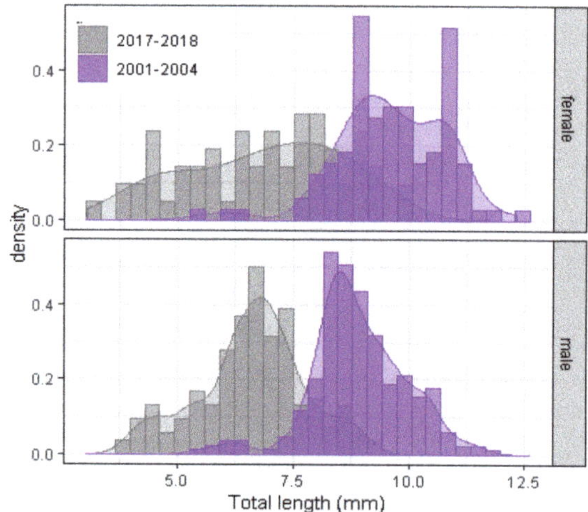


**Figure 3.** Comparison between 2017–2018 (grey) and 2001–2004 (purple) data [22] on body length of females (top) and males (bottom) of *Orconectes limosus* in Lake Maggiore. Histograms represent the observed values, whereas the smoothed line is the kernel density estimate of the distribution.

In Spring 2018 (betwen 27th April and 30th May), ten ovigerous females were captured in two sites (Dormelletto and Lisanza) in four different sampling sessions. Females measured between 50.4 mm and 78.6 mm (mean number of eggs/female: 72)

### 3.3. Influence of Environmental Factors on O. limosus Distribution

Following data exploration [34], we dropped from the regression analysis the variable average precipitation, warmer temperature, and colder temperature, being mutually collinear ($|r| > 0.7$) and also associated with the categorical variables of shelter presence. No outliers were present in the dataset. Coplot revealed a potential interaction between mean annual temperature and the presence of shelters, which we incorporated in the initial regression structure.

The initial model included substrate type (Substrate), presence of algae (Algae), presence of macrophytes (Macrophytes), and presence of shelters in interaction with mean annual temperature (Shelters * Tmean).

According to model selection (Table 3), the most appropriate model structure supported by the observations explaining the habitat preference of the alien crayfish had the following structure:

$$y \sim \text{Tmean} * \text{Shelters} \tag{1}$$

**Table 3.** Model selection according to the corrected Akaike information criterion for finite sample size (AICc [35]). Model are ordered from the least to the most appropriate. df: degrees of freedom, ΔAICc: difference of AICc, wi (AIC): Rounded Akaike weights *sensu* Burnham & Anderson [33].

| Model Structure | Df | AICc | ΔAICc | wi(AIC) |
|---|---|---|---|---|
| y ~ Tmean * Shelters + Substrate + Algae + Macrophytes | 8 | 74.00 | 5.27 | 0.02 |
| y ~ Tmean * Shelters + Algae + Macrophytes | 6 | 70.27 | 1.54 | 0.18 |
| y ~ Tmean * Shelters + Macrophytes | 5 | 68.81 | 0.08 | 0.38 |
| y ~ Tmean * Shelters | 4 | 68.73 | 0.00 | 0.40 |

The significant interaction (Tmean * Shelters, estimated $\beta \pm$ s.e. $= -6.66 \pm 3.02$, $p = 0.02$) reflects a differential response to mean annual temperature (Tmean $= 4.89 \pm 2.23$, $p = 0.02$) depending on the presence or absence of shelters (Shelters = 80.97 ± 36.09, $p$ = 0.02). The probability of presence of the species increased positively with increasing mean annual temperature in habitats lacking shelters, whereas the trend was flat to slightly negative in areas with shelters (Figure 4).

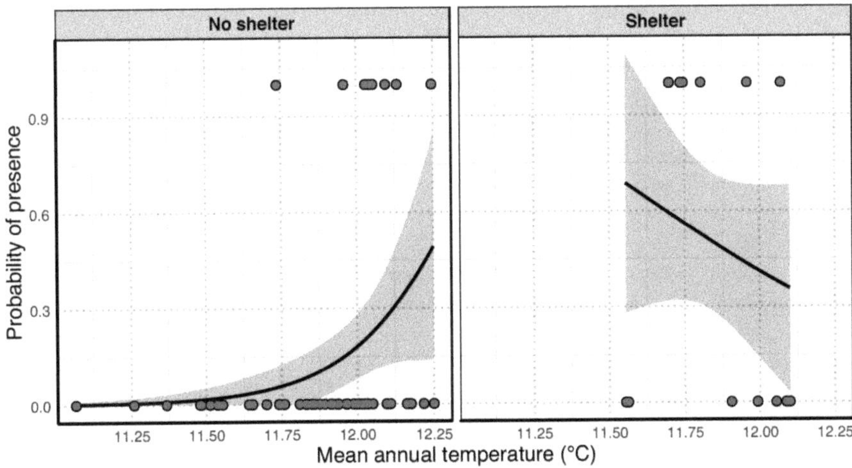

**Figure 4.** Predicted relationships between the presence-absence of *Orconectes limosus* and the mean annual temperature in interaction with the presence of shelters. Shaded grey surfaces are 95% confidence intervals. Blue dots are observed values.

## 4. Discussion

We documented the occurrence of three alien crayfish species in the deep subalpine Lake Maggiore: *Pacifastacus leniusculus*, *Procambarus clarkii*, and *Orconectes limosus*. Concerning the first species, this is the first record in Lake Maggiore. In Switzerland, *P. leniusculus* was introduced around 1980, although it showed a scattered distribution [40]. In 2007, one specimen was also found in Canton Tessin, in a tributary of the lake near Minusio [41]. These authors hypothesized that the finding was linked to an occasional introduction, and argued that the species was still not of concern. The specimens from our study were found in 2015 (one individual) and in 2017 (two individuals), and were reported by a local fisherman in the lake zone beyond Tenero/Mappo and Rivapiana of Minusio. This is suggestive of an on-going process of acclimatization of the species on the littoral shores of the lake standing Minusio's river tributary. However, as low water temperature seems to be the major predictor that determines the abundance of the signal crayfish (12.7 °C) [42], the cold waters of the lake recorded during winter seasons (average winter temperature 3.9 °C, data available at www.cipais.org could prevent, or slow down, its spread. Further analyses are needed to clarify this aspect.

The species *P. clarkii* has been regularly found in Lake Maggiore since 2016, although records are only anecdotal [21]. This species is recognized as the most successful IAS in Italy [43]. *Procambarus clarkii* is considered a typical inhabitant of warm waters, being tolerant to water eutrophication and mineralization [44]. Notwithstanding suitable environmental features for its establishment are present in the lake, and the fact that a stable population have been reported for the nearby Lake Orta [45], our data suggests that the species is not widespread in Lake Maggiore yet. Delmastro [21] reported observations from researchers and operators of the Environmental Agencies during monitoring activities in the lake, but its occurrence is seemingly linked to the presence of waterways, such as tributaries or outlet of the lake. Local fishermen, who were interviewed, confirmed this tendency. At present, no record is available for the lake side representing the Verbano-Cusio-Ossola province.

A thorough population analysis was possible only for *O. limosus*. In Italy, the species in known since 1991 [46], when it was accidentally introduced from Poland. Currently, it is widespread particularly in the Po Plain [43]. In Lake Maggiore, it was firstly reported by Bazzoni [22], who found specimens in the Borromean Gulf on the central part of the lake. Our results demonstrate that *O. limosus* occurs mainly on the central-southern part of the lake, near the River Ticino outlet connecting the lake to the Po Plain. In a previous paper, focused on the spatial dynamic of invasion and on the establishment of the invasive bivalves *Corbicula fluminea* Müller, 1774 in the lake, Kamburska and coauthors [20] hypothesized that this mollusc initially settled in the southern basin of the lake, starting from populations established in the River Ticino outlet [47]. Quite possibly, we are observing a similar invasion dynamic, i.e., the spread of the species from the River Ticino lake-outlet to the southern and northern district of the lake. Bazzoni [22] reported that the population was changing during the monitoring period, observing a decrease in the number of individuals in the Fondotoce Nature Reserve from 2000 to 2004. In the Nature Reserve, we never detected the species, both through visual census or trapping sessions in two consecutive years. Tentatively, we suggest that its decrease in the area is linked to the presence of a large Grey Heron (*Ardea cinerea*) colony near the Reserve (Figure 5). Indeed, terrestrial predators such as herons, belonging to *Butorides* and *Ardea* genera, have been demonstrated to caused little crayfish mortality in deep areas, but can rapidly consume both small and large crayfish exposed in shallow areas [48]. Unfortunately, the heronry is outside the Nature Reserve, on the opposite side of a nearby congested road that is periodically subjected to side tree cutting. Further studies are required to understand the effectiveness of bird predation on the population of *O. limous*. However, since these ardeids (species included in the Bird Directive 2009/147/EC) can consume a large amount of crayfish, we suggest that the heronry of the Fondotoce Nature Reserve to be included in under the umbrella of protection of the Site of Community Importance (SCI).

**Figure 5.** The heronry (**white arrow**) near the Fondotoce Nature Reserve (**black arrow**) in Lake Maggiore.

Concerning biometrics, we did not find a significant difference between males and females for total body length and total weight. Although sexual dimorphism is widely diffused in Decapods [49], this has not been fully demonstrated for *O. limosus*. Pieplow [29] did not distinguished between males and females in total body length. Dǔriš and coauthors [50] found a difference in the total body length between sexes, but males were larger than females in brooks and isolated waters, and smaller in large rivers. Pilotto and coauthors [51], who observed a similar biometry in *O. limosus* populations in the nearby Lake Varese, did not found significant differences in cephalotorax length between sexes, observing differences only within age classes. These differences in findings are probably related to a unique feature of the females of this species: similarly to the conspecific males, they undergo significant cyclic changes in size, producing larger chelae, abdomen, and body dimensions, especially during the moult to form I (i.e., when an adult crayfish changes from a sexual inactive to a sexual active form) [52].

Differences in biometry and size classes, between present and past results available for the Lake Maggiore [22], can be linked to the different trapping method employed in the present time respect to the past (2001–2004). We employed traps with a mesh size of 1 $cm^2$, allowing us to catch also young-of-the-year individuals, while Bazzoni [22] employed traps with meshes of 4–5 cm, unsuitable for catching small-sized individuals (Figure 3).

Concerning the environmental factors affecting crayfish distribution, we found that, in the absence of shelters, the probability of occurrence of *O. limosus* increases with the raise of mean annual temperatures. This result suggests that the species succeeds in warmer areas, while in presence of refugees, there is not a specific selection of areas to be colonized. Thus, the lack of suitable refuges may be a strong limiting factor for the species spread in absence of optimal climatic conditions, as confirmed for the species *P. leniusculus* [53]. This result is in accordance with general studies on the ecological drivers of invasive crustaceans, such as crayfish [53–55] and crabs [56], pointing out the crucial role of climate in determining the successful outcome of biological invasions.

It is interesting to note that we did not found any specimen in the northern part of the lake. The substrate along the coasts changes along the south-north axes from sandy beaches, mixed with gravel and algae formations in the southern part, into rocky cliffs in the central part of the lake, to beaches with boulders in the northern part. The boulders were settled in by humans to create narrow beaches for tourists and as docking sites for small boats. However, the substrate was not a significant factor in our regression model, and was dropped through model selection. This result contrasts with previous observations on other crayfish species, where significant relationships between granulometry and species abundance were observed [42,57].

The finding of few ovigerous females in April and May 2018 indicate the Lake Maggiore as a potentially suitable reproductive site for the species. Furthermore, according to the structure of age/size classes based on the classic work of Pieplow [29], females were between the first and the second year of life. This is contrasting to previous observations [58,59], which suggested that females reach sexual maturity from the second year of life.

From a methodological point of view, we support the efficiency of visual inspection for the detection of crayfish [25,60] to be used as a complementary tool of traps. By setting a standard time for preliminary inspection through visual survey, we were able to verify 75 sampling points for a total of 170 km of inspected shoreline in a reasonable time. Moreover, only through inspections we were able to assess the occurrence of *P. clarkii*, as this species never felt into traps. Although the method is not without bias [25], visual encounter survey is an effective and inexpensive approach to screen for the presence of alien crayfish from large lakes.

## 5. Conclusions

The Invasive Alien Species in Lake Maggiore monitoring project was a pilot study that allowed us to detect the occurrence of three alien species in this large subalpine lake, and to shed some light on the autoecology of the most abundant species, *O. limosus*. The recovery of these species of Union Concern raises important conservation issues that need to be addressed. The finding of *P. leniusculus* in the Swiss part of the lake is worrisome because of the proximity of the Bolle di Magadino Nature Reserve, and of several source population of the native *Austropotamobius pallipes* in the lake catchment. Moreover, the alien species, observed in this study, not only compete with the native fauna, but are also the primary vectors of the oomycetes *Aphanomyces astaci* Schikora 1906, the causative agent of the crayfish plague [61]. We suggest that Italy and Switzerland should agree on concrete actions to prevent the spread of crayfish by intensifying trapping sessions in the spring reproductive period. Further investigations are needed to understand the ecology and the behavior of invasive crayfish, in particular, concerning their vertical distribution in relation to depth and the possible interaction with natural predators such as fish and ardeids. Altogether, the evidence may contribute to reducing the level of uncertainty of environmental management actions to be proposed at the catchment scale.

**Author Contributions:** Conceptualization, L.G. and A.B.; data curation, L.G. and M.C.; formal analysis, L.G. and S.M.; funding acquisition, A.B.; investigation, L.G. and A.B.; methodology, L.G. and A.B.; project administration, A.B.; software, S.M.; visualization, L.G. and S.M.; writing—original draft, L.G., S.M., M.C. and A.B. All authors have read and agreed to the published version of the manuscript.

**Funding:** The present study was funded by the International Commission for the Protection of Italian-Swiss Waters (CIPAIS), and by ARPA Lombardia (Regional Agency for Environmental Protection).

**Acknowledgments:** We are thankful to Nicola Patocchi and Davide Sargenti (Fondazione Bolle di Magadino), Luca Migliori, Leonardo Mantovani, two students of the high schools of Omegna and Verbania Alessandro Poletti, Tommaso Zaneboni, and Marta Dugaro for their technical support during samplings. We are grateful to Walter Branca for providing *P. leniusculus* individuals examined in this study. Thanks are also due to Paolo Bazzoni (Ossolana Acque) for sharing the results of his previous work. We gratefully acknowledge two anonymous reviewers for the time and effort spent in reviewing our article and ensuring a high quality publication. Catch permits were requested and obtained from: Commissariato italiano per la Convenzione italo-svizzera sulla Pesca; Ente di gestione delle aree protette del Ticino e del Lago Maggiore; Parco Lombardo del Ticino, Repubblica e Canton Ticino-Ufficio della natura e del paesaggio (Switzerland); Fondazione Bolle di Magadino (Switzerland); Provincia di Novara; Provincia di Varese; Provincia del VCO; Polizia Provinciale di Varese; and Polizia Provinciale VCO. The Ufficio Federale dell'Ambiente (UFAM), Divisione Specie, ecosistemi e paesaggi (Switzerland) has also given us the permission of export live specimens of alien crayfish for scientific purposes.

**Conflicts of Interest:** The authors declare no conflict of interest.

## References

1.  Butchart, S.H.; Walpole, M.; Collen, B.; Van Strien, A.; Scharlemann, J.P.; Almond, R.E.; Baillie, J.E.M.; Bomhard, B.; Brown, C.; Bruno, J.; et al. Global biodiversity: Indicators of recent declines. *Science* **2010**, *328*, 1164–1168. [CrossRef]

2.  Ehrenfeld, J.G. Ecosystem consequences of biological invasions. *Annu. Rev. Ecol. Evol. Syst.* **2010**, *41*, 59–80. [CrossRef]

3.  Bellard, C.; Cassey, P.; Blackburn, T.M. Alien species as a driver of recent extinctions. *Biol. Lett.* **2016**, *12*, 20150623. [CrossRef]

4.  IPBES 2019. Global Assessment Report on Biodiversity and Ecosystem Services of the Intergovernmental Science-Policy Platform on Biodiversity and Ecosystem Services. Available online: www.ipbes.net (accessed on 9 September 2019).

5.  Westman, K. Alien crayfish in Europe: Negative and positive impacts and interactions with native crayfish. In *Invasive Aquatic Species of Europe. Distribution, Impacts and Management*; Leppäkoski, E., Gollasch, S., Olenin, S., Eds.; Springer: Dordrecht, The Netherlands, 2002.

6.  Gherardi, F. *Crayfish in Europe as Alien Species*; CRC Press: Boca Raton, FL, USA; Routledge: London, UK, 1999; p. 310.

7.  Cantonati, M.; Poikane, S.; Pringle, C.M.; Stevens, L.E.; Turak, E.; Heino, J.; Richardson, J.S.; Bolpagni, R.; Borrini, A.; Cid, N.; et al. Characteristics, main impacts, and stewardship of natural and artificial freshwater environments: Consequences for biodiversity conservation. *Water* **2020**, *12*, 260. [CrossRef]

8.  Solimini, A.; Cardoso, A.C.; Heiskanen, A.S. Indicators and methods for the Ecological Status Assessment under the Water Framework Directive. In *Linkages between Chemical and Biological Quality of Surface Waters*; EUR 22314 EN; European Commission: Rome, Italy, 2006; p. 248.

9.  Nõges, P.; van de Bund, W.; Cardoso, A.C.; Heiskanen, A.S. Impact of climatic variability on parameters used in typology and ecological quality assessment of surface waters—Implications on the Water Framework Directive. *Hydrobiologia* **2007**, *584*, 373–379. [CrossRef]

10. Gherardi, F.; Aquiloni, L.; Diéguez-Uribeondo, J.; Tricarico, E. Managing invasive crayfish: Is there a hope? *Aquat. Sci.* **2011**, *73*, 185–200. [CrossRef]

11. Chucholl, C.; Daudey, T. First record of *Orconectes juvenilis* (Hagen, 1870) in eastern France: Update to the species identity of a recently introduced orconectid crayfish (Crustacea: Astacida). *Aquat. Invasions* **2008**, *3*, 105–107. [CrossRef]

12. EU 2017 Invasive Alien Species of Union Concern. Available online: https://ec.europa.eu/environment/nature/pdf/IAS_brochure_species.pdf (accessed on 27 March 2020).

13. Boggero, A.; Basset, A.; Austoni, M.; Barbone, E.; Bartolozzi, L.; Bertani, I.; Campanaro, A.; Cattaneo, A.; Cianferoni, F.; Corriero, G.; et al. Weak effects of habitat type on susceptibility to invasive freshwater species: An Italian case study. *Aquat. Conserv.* **2014**, *24*, 841–852. [CrossRef]

14. Colangelo, P.; Fontaneto, D.; Marchetto, A.; Ludovisi, A.; Basset, A.; Bartolozzi, L.; Bertani, I.; Campanaro, A.; Cattaneo, A.; Cianferoni, F.; et al. Alien species in Italian freshwater ecosystems: A macroecological assessment of invasion drivers. *Aquat. Invasions* **2017**, *12*, 299–309. [CrossRef]

15. Guilizzoni, P.; Levine, S.N.; Manca, M.; Marchetto, A.; Lami, A.; Ambrosetti, W.; Brauer, A.; Gerli, S.; Carrara, E.A.; Rolla, A.; et al. Ecological effects of multiple stressors on a deep lake (Lago Maggiore, Italy) integrating neo and palaeolimnological approaches. *J. Limnol.* **2012**, *71*, 1–22. [CrossRef]

16. Callieri, C.; Piscia, R. Photosynthetic efficiency and seasonality of autotrophic picoplankton in Lago Maggiore after its recovery. *Freshw. Biol.* **2002**, *47*, 941–956. [CrossRef]

17. Mosello, R.; Lami, A. Climate Change and Related Effects on Water Quality: Examples from Lake Maggiore (Italy). *Glob. Bioeth.* **2011**, *24*, 95–98. [CrossRef]

18. Volta, P.; Jepsen, N. The recent invasion of *Rutilus rutilus* (L.)(Pisces: Cyprinidae) in a large South-Alpine lake: Lago Maggiore. *J. Limnol.* **2008**, *67*, 163–170. [CrossRef]

19. Fenoglio, S.; Bo, T.; Cucco, M.; Mercalli, L.; Malacarne, G. Effects of global climate change on freshwater biota: A review with special emphasis on the Italian situation. *Ital. J. Zool.* **2010**, *77*, 374–383. [CrossRef]

20. Kamburska, L.; Lauceri, R.; Beltrami, M.; Boggero, A.; Cardeccia, A.; Guarneri, I.; Manca, M.; Riccardi, N. Establishment of *Corbicula fluminea* (OF Müller, 1774) in Lake Maggiore: A spatial approach to trace the invasion dynamics. *Bioinvasions Rec.* **2013**, *2*, 105–117. [CrossRef]

21. Delmastro, G.B. Il gambero della Louisiana *Procambarus clarkii* (Girard, 1852) in Piemonte: Nuove osservazioni su distribuzione, biologia, impatto e utilizzo (Crustacea: Decapoda: Cambaridae). *Rivista Piemontese Di Storia Naturale* **2017**, *38*, 61–129.

22. Bazzoni, P. *Censimento E Studio Delle Popolazioni Di Gambero d'Acqua Dolce Nell'Area Del Verbano-Cusio-Ossola*; Azienda Agricola Ossolana Acque: Verbano, Italy, 2006; p. 25.

23. Saidi, H.; Ciampittiello, M.; Dresti, C.; Ghiglieri, G. Assessment of trends in extreme precipitation events: A case study in Piedmont (North-West Italy). *Water Resour. Manage.* **2015**, *29*, 63–80. [CrossRef]

24. Bain, M.B.; Stevenson, N.J. Aquatic habitat assessment. In *Asian Fisheries Society*; Bethesda: Rockville, MD, USA, 1999.

25. Bonk, M.; Bobrek, R.; Dołęga, J.; Strużyński, W. Evaluation of visual encounter surveys of the noble crayfish, *Astacus astacus*, and the spiny-cheek crayfish, *Orconectes limosus*. *Fish. Aquat. Life* **2019**, *27*, 112–117. [CrossRef]

26. Tricarico, E.; Inghilesi, A.F.; Ferretti, G.; Johovic, I.; Ruberti, V.; Scapini, F. *Monitoraggio e quarto intervento di controllo del gambero rosso della Louisiana Procambarus clarkii*; Quinta relazione tecnica A2C2 per il progetto LIFE+ 2011 SOS TUSCAN WETLANDS; Italy, 2017; Available online: http://www.life-sostuscanwetlands.eu/index.php/it/pubblicazioni/ (accessed on 1 March 2020).

27. Boggero, A.; Dugaro, M.; Migliori, L.; Garzoli, L. Prima segnalazione del gambero invasivo Pacifastacus leniusculus (Dana 1852) nel Lago Maggiore (Cantone Ticino, Svizzera). *Bollettino Della Società Ticinese Di Scienze Naturali* **2018**, *106*, 103–106.

28. Souty-Grosset, C.; Holdich, D.M.; Noël, P.Y.; Reynolds, J.D.; Haffner, P. *Atlas of Crayfish in Europe*; Muséum national d'Histoire naturelle: Paris, France, 2006; p. 187.

29. Pieplow, U. *Fischereiwissenschaftliche Monographie Von Cambarus Affinis*; Say. Neumann: Berlin, Germany, 1939; p. 92.

30. *R Core Team R: A language and environment for statistical computing*; R Foundation for Statistical Computing: Vienna, Austria, 2013. Available online: https://www.R-project.org/ (accessed on 13 May 2020).

31. Wickham, H.; Chang, W.; Wickham, M.H. Package 'ggplot2'. In *Create Elegant Data Visualisations Using the Grammar of Graphics*; 2016; Volume 2, pp. 1–189. Available online: https://ggplot2.tidyverse.org/ (accessed on 13 May 2020).

32. Fick, S.E.; Hijmans, R.J. WorldClim 2: New 1-km spatial resolution climate surfaces for global land areas. *Int. J. Climatol.* **2017**, *37*, 4302–4315. [CrossRef]

33. Zuur, A.F.; Ieno, E.N. A protocol for conducting and presenting results of regression-type analyses. *Methods Ecolo. Evol.* **2016**, *7*, 636–645. [CrossRef]

34. Zuur, A.F.; Ieno, E.N.; Elphick, S.C. A protocol for data exploration to avoid common statistical problem. *Methods Ecol. Evol.* **2010**, *1*, 3–14. [CrossRef]

35. Zuur, A.F.; Ieno, E.N.; Walker, N.J.; Savaliev, A.A.; Smith, G.M. *Mixed Effect Models and Extensions in Ecology with R.*; Springer: Berlin, Germany, 2009; p. 574.

36. Burnham, K.P.; Anderson, D.R. *Model Selection and Multimodel Inference: A Practical Information-Theoretic Approach*; Springer: New York, NY, USA, 2002; p. 488.

37. Johnson, J.B.; Omland, K.S. Model selection in ecology and evolution. *Trends Ecol. Evol.* **2004**, *19*, 101–108. [CrossRef] [PubMed]

38. Hurvich, C.M.; Tsai, C.L. Regression and time series model selection in small samples. *Biometrika* **1989**, *76*, 297–307. [CrossRef]

39. Howkins, D.M. The problem of overfitting. *J. Chem. Inf. Comput. Sci.* **2004**, *44*, 1–12. [CrossRef]

40. Hefti, D.; Stucki, P. Crayfish management for Swiss waters. *Bulletin Français De La Pêche Et De La Pisciculture* **2006**, *380–381*, 937–950. [CrossRef]

41. Maddalena, T.; Zanini, M.; Torriani, D.; Marchesi, P.; Jann, B.; Paltrinieri, L. Inventario dei gamberi d'acqua dolce del Cantone Ticino (Svizzera). *Bollettino Della Società Ticinese Di Scienze Naturali* **2009**, *97*, 19–25.

42. Vedia, I.; Galicia, D.; Baquero, E.; Oscoz, J.; Miranda, R. Environmental factors influencing the distribution and abundance of the introduced signal crayfish in the north of Iberian Peninsula. *Mar. Freshw. Res.* **2017**, *68*, 900–908. [CrossRef]

43. Aquiloni, L.; Tricarico, E.; Gherardi, F. Crayfish in Italy: Distribution, threats and management. *Int. Aquat. Res.* **2010**, *2*, 1–14.

44. Maceda-Veiga, A.; De Sostoa, A.; Sanchez-Espada, S. Factors affecting the establishment of the invasive crayfish *Procambarus clarkii* (Crustacea, Decapoda) in the Mediterranean rivers of the northeastern Iberian Peninsula. *Hydrobiologia* **2013**, *703*, 33–45. [CrossRef]

45. Piscia, R.; Volta, P.; Boggero, A.; Manca, M. The invasion of Lake Orta (Italy) by the red swamp crayfish *Procambarus clarkii* (Girard, 1852): A new threat to an unstable environment. *Aquat. Invasions* **2011**, *6*, S45–S48. [CrossRef]

46. Gherardi, F.; Baldaccini, G.N.; Barbaresi, S.; Ercolini, P.; De Luise, G.; Mazzoni, D.; Mori, M. The situation in Italy. In *Crayfish in Europe as Alien Species. How to Make the Best of a Bad Situation?* Gherardi, F., Holdich, D.M., Balkema, A.A., Eds.; Crustacean Issues: Rotterdam, The Netherlands, 1999; pp. 107–128.

47. Groppali, R Sulla presenza del gambero americano *Orconectes limosus* (Rafinesque) in acque della Pianura Pavese (Crustacea Decapoda Cambaridae). *Riv. Piem. St. Nat.* **1993**, *14*, 93–96.

48. Englund, G.; Krupa, J.J. Habitat use by crayfish in stream pools: Influence of predators, depth and body size. *Freshw. Biol.* **2000**, *43*, 75–83. [CrossRef]

49. Vlach, P.; Valdmanová, L. Morphometry of the stone crayfish (*Austropotamobius torrentium*) in the Czech Republic: Allometry and sexual dimorphism. *Knowl. Manag. Aquat. Ec.* **2015**, *416*, 16–50. [CrossRef]

50. Ďuriš, Z.; Drozd, P.; Horká, I.; Kozák, P.; Policar, T. Biometry and demography of the invasive crayfish Orconectes limosus in the Czech Republic. *Bulletin Français De La Pêche Et De La Pisciculture* **2006**, *380–381*, 1215–1228. [CrossRef]

51. Pilotto, F.; Free, G.; Crosa, G.; Sena, F.; Ghiani, M.; Cardoso, A.C. The invasive crayfish *Orconectes limosus* in Lake Varese: Estimating abundance and population size structure in the context of habitat and methodological constraints. *J. Crustacean Biol.* **2008**, *28*, 633–640. [CrossRef]

52. Buřič, M.; Kouba, A.; Kozák, P. Intra-sex dimorphism in crayfish females. *Zoology* **2010**, *113*, 301–307. [CrossRef]

53. Usio, N.; Nakajima, H.; Kamiyama, R.; Wakana, I.; Hiruta, S.; Takamura, N. Predicting the distribution of invasive crayfish (*Pacifastacus leniusculus*) in a Kusiro Moor marsh (Japan) using classification and regression trees. *Ecol. Res.* **2006**, *21*, 271–277. [CrossRef]

54. Liu, X.; Guo, Z.; Ke, Z.; Wang, S.; Li, Y. Increasing potential risk of a global aquatic invader in Europe in contrast to other continents under future climate change. *PLoS ONE* **2011**, *6*, e18429. [CrossRef] [PubMed]

55. Zhang, Z.; Capinha, C.; Usio, N.; Weterings, R.; Liu, X.; Li, Y.; Landeira, J.; Zhou, Q.; Yokota, M. Impacts of climate change on the global potential distribution of two notorious invasive crayfishes. *Freshw. Biol.* **2020**, *65*, 353–365. [CrossRef]

56. Zhang, Z.; Mammola, S.; McLay, C.L.; Capinha, C.; Yokota, M. To invade or not to invade? Exploring the niche-based processes underlying the failure of a biological invasion using the invasive Chinese mitten crab. *Sci. Total Environ.* **2020**, *728*, 138815. [CrossRef] [PubMed]

57. Jowett, I.G.; Parkyn, S.M.; Richardson, J. Habitat characteristics of crayfish (*Paranephrops planifrons*) in New Zealand streams using generalised additive models (GAMs). *Hydrobiologia* **2008**, *596*, 353–365. [CrossRef]

58. Holdich, D.; Black, J. The spiny-cheek crayfish, *Orconectes limosus* (Rafinesque, 1817)Crustacea: Decapoda: Cambaridae., digs into the UK. *Aquat. Invasions* **2007**, *2*, 1–15. [CrossRef]

59. Momot, W.T. Orconectes in North America and elsewhere. In *Freshwater Crayfish: Biology, Management and Exploitation*; Holdich, D.M., Lowery, R.S., Eds.; Croom Helm Ltd.: Kent, UK, 1998; pp. 262–283.

60. Fulton, C.J.; Starrs, D.; Ruibal, M.P.; Ebner, B.C. Counting crayfish: Active searching and baited cameras trump conventional hoop netting in detecting *Euastacus armatus*. *Endanger Species Res.* **2012**, *19*, 39–45. [CrossRef]

61. Alderman, D.J.; Polglase, J.L. *Aphanomyces astaci*: Isolation and culture. *J. Fish Dis.* **1986**, *9*, 367–379. [CrossRef]

*Article*

# Selection of Macroinvertebrate Indices and Metrics for Assessing Sediment Quality in the St. Lawrence River (QC, Canada)

Mélanie Desrosiers [1,*], Bernadette Pinel-Alloul [2,*] and Charlotte Spilmont [3]

[1] Division de l'Écotoxicologie et de l'Évaluation du Risque, Centre d'Expertise en Analyse Environnementale du Québec, Ministère de l'Environnement et de la Lutte Contre les Changements Climatiques (MELCC), Québec, QC G1P 3W8, Canada

[2] Groupe de Recherche Interuniversitaire en Limnologie (GRIL), Département de Sciences Biologiques, Université de Montréal, Montréal, QC H2V 0B3, Canada

[3] UFR Sciences fondamentales et appliquées, Université de Lorraine, 57070 Metz, France; charlotte.spilmont@laposte.fr

* Correspondence: melanie.desrosiers@environnement.gouv.qc.ca (M.D.); bernadette.pinel-alloul@umontreal.ca (B.P.-A.)

Received: 5 November 2020; Accepted: 25 November 2020; Published: 27 November 2020

**Abstract:** This study aims to evaluate the anthropogenic pressure in the St. Lawrence River by assessing the relationships between composition and chemical contamination of sediments and macroinvertebrate community structure using a selection of indices and metrics. The aims of this study are to (i) determine the composition of macroinvertebrate community in sediments across a gradient of disturbance, (ii) select relevant macroinvertebrate indices and metrics for the assessment of sediment quality, (iii) investigate whether responses of selected indices and metrics differ across habitats and/or sediment quality classes, and finally, (iv) determine the thresholds for critical contaminants related to significant changes in the most relevant indices and metrics. Organic and inorganic contaminants as well as other sediment variables (sediment grain size, total organic carbon, nutrients, etc.) and macroinvertebrate assemblages were determined in 59 sites along the river. Fourteen macroinvertebrate indices and metrics, on the 264 initially selected, were shown to be the most effective to be used in bioassessment for the St. Lawrence River. However, the variation in macroinvertebrate indices and metrics remains strongly explained by habitat characteristics, such as sediment grain size or the level of nutrients. There is also an influence of metals and, to a lesser extent, organic contaminants such as petroleum hydrocarbons. The 14 selected indices and metrics are promising bioassessment tools that are easy to use and interpret in an environmental assessment of sediment quality in the St. Lawrence River.

**Keywords:** bioassessment; macroinvertebrates; indices and metrics; sediment quality; St. Lawrence River

---

## 1. Introduction

According to the requirements of the European Water Framework Directive [1,2], the Canadian Aquatic Biomonitoring Network (CABIN; [3]), and the U.S. Environmental Protection Agency [4], macroinvertebrates have been commonly used for the bioassessment of anthropogenic disturbances in rivers because they are (i) reliable bioindicators of water and sediment qualities [5,6], (ii) efficient and cost-effective biomonitoring tools [7,8], and (iii) useful to differentiate reference conditions from impaired sites [9,10]. In streams and small rivers, studies showed that the structure of the benthic macroinvertebrate community reflects the impacts of anthropogenic disturbances such as water acidification, organic pollution, metal contamination, and habitat degradation [11–14]. In large rivers, changes in the macroinvertebrate community are related to multiple environmental factors [15], including changes in habitat vegetation [16,17], water-level fluctuations [18], water quality [19,20], sediment grain size and contamination [21–23], and human disturbances [24].

Bioassessment approaches comparing reference and disturbed sites are designed to determine whether poor water or sediment qualities are stressing the macroinvertebrate community beyond the range of natural variation [25,26]. However, this is a difficult task due to the complexity of river ecosystems and the interaction of multiple factors which limit the possibility to predict the overall responses of macroinvertebrate assemblages to environmental changes, either natural or anthropogenic. Difficulty to find reference sites in large and complex rivers under significant anthropogenic stressors is another problematic issue that further limits bioassessment. Approaches comparing sites on a disturbance gradient are now more suitable for bioassessment in rivers because they help to establish the relationships between macroinvertebrate community structure and natural environmental conditions and anthropogenic stressors [6,23].

Over the past 20 years, 60% of the biological indicators used to assess ecological quality of rivers were based on macroinvertebrate communities [27]. A myriad of indices and metrics have been applied in bioassessment approaches to establish the various sensitivities of macroinvertebrates to different types of disturbances. However, the development of most suitable macrobenthic indices and metrics for the bioassessment of the ecological quality status of rivers and lakes is still in progress. The first indices that come to mind are the diversity indices and metrics based on taxon richness, used since the eighties. However, their relevance has been discussed [28] because taxon richness and diversity indices depend more on geographical, climatic, historical, and ecological factors than on the direct impact of anthropogenic stressors, except in the case of extreme physical or chemical disturbance. Diversity indices and metrics alone are no longer recognized as relevant tools in biological assessment [29] but they could be included such as multimetric STAR-ICM index in small, lowland rivers in Europe [30]. Biotic indices combining richness and abundance of sensitive or tolerant taxa were more successful in detecting ecological changes among sites and the effects of anthropogenic stressors in rivers [31], lakes [32] and ponds [33]. Among the indices used to assess stressor-specific disturbances, we can cite: (i) the saprobic index [8,34] and the Hilsenhoff index [35,36] for organic pollution, (ii) the Index of Community Sensitivity (ICS, [22]) or the Invertebrate Community Index (ICI, [37]) for water quality and sediment metal contamination. More complex integrated monitoring based on multimetric procedures has recently been implemented for biological assessment in Europe [12,20,38,39] and North America [22,24,37,40]. These procedures allow the selection and aggregation of metric scores in a single index that helps to determine whether action or restoration is needed and simplify management and decision-making. Successful multimetric score procedures have been validated for river pollution surveys. As examples, we can cite: (i) the Biological Monitoring Working Party (BMWP) used in Spain [41], the UK [12], Poland [27], and Canada [42], (ii) the Index of Biotic Integrity (IBI, [43]) and the Panel Index [40] used in the USA, (iii) the Belgian Biotic Index (BBI, [44]) or the Multimetric Macroinvertebrate Index Flanders (MMIF, [12]) used in Belgium, and (iv) the Macroinvertebrate-Based Multimetric Index (IBMA) applied in Martinique and Guadeloupe territories [39]. In Canada, multimetric indices have been developed to assess sensitivity of benthic biota to river flow regimes [45] and water quality [42]. However, in Canada, there are still few developments towards a multimetric approach using macroinvertebrates to assess sediment contamination in large rivers compared to those in other countries [15,24,46,47]. The future needs for the development of sediment bioassessment methods in large rivers include: (i) the selection of relevant macroinvertebrates indices and metrics based on ecological principles underlying metric choice for specific disturbances, (ii) the validation of the potential of indices and metrics to discriminate sites according to a gradient of environmental conditions and disturbances, and (iii) the determination of criteria and management thresholds that indicate environmental degradation and the need for quantitative assessment studies and remediation projects.

The present research focuses on the St. Lawrence River (QC, Canada), one of the most important large rivers in the world draining a watershed area of 1,610,000 km$^2$ and flowing across 1000 km from Lake Ontario to the Gulf of St. Lawrence. Since the 1950s, intensive agriculture, urbanization, and industrialization have caused the contamination of sediments in the St. Lawrence River [48]. Macroinvertebrates are a critical component of the wetland and sediment food webs in the St. Lawrence River [18]. Previous studies showed that their distribution, community composition and functional traits vary according to ecological and toxicological factors such as vegetation types (filamentous algal mats, emergent and submerged macrophytes) [16], water quality and sediment contamination [17,23,37,49], landscape features and hydrological regime [18]. However, most of these studies were limited to littoral wetland habitats or to specific approaches based on species assemblages and functional traits. Since macroinvertebrates can be impacted by sediment contaminants, their biomonitoring is required to complete the contamination assessment, to evaluate the ecotoxicological risk, and to determine the remediation needs at sediment-contaminated sites in the St. Lawrence River. This large river is also an essential transportation route in northeastern America, and periodic dredging of sediment is required for the maintenance of the waterway and harbour facility. Because dredged sediments may contain a range of contaminants that could affect benthic organisms at deposit sites, environmental risk assessment and management of these dredging projects are required to determine sediment quality [50].

No study has yet compared the potential of multiple indices and metrics to assess specific stressors such as sediment dredging and contamination across the fluvial section of the St. Lawrence River. The objectives of this study are as follows: (i) determine the composition of macroinvertebrate community in sediments of typical habitats across a gradient of disturbance, (ii) select relevant indices and metrics from a panel of macroinvertebrate indices and metrics based on their ecological relevance for the assessment of sediment quality and contamination and on their potential for large river ecotoxicological risk assessment, (iii) investigate whether sensitivity of selected indices and metrics differ across habitats and/or sediment quality classes, and finally, (iv) determine the thresholds for critical contaminants related to significant changes in the most relevant indices and metrics.

## 2. Materials and Methods

### 2.1. Sampling Sites at the St. Lawrence River

The study covers a 240 km longitudinal transect of the fluvial section of the St. Lawrence River including three fluvial lakes (Lake Saint-François, Lake Saint-Louis, and Lake Saint-Pierre) and the Montreal Harbour area (Figure 1). Sampling sites were in sedimentation zones impacted by fine-particle deposition, potential dredging, and past or present point sources of anthropogenic contamination [23]. A total of 59 sites were distributed among the different habitat zones and sampled during the fall of 2004 and 2005, because it is the period of higher biomass in the St. Lawrence River [17,23]. Ten sites were sampled in Lake Saint-François (LSF), the first natural enlargement of the St. Lawrence River downstream of Lake Ontario. This fluvial lake is relatively oligotrophic, shallow and covered with submerged macrophytes over most of its western section [37]. Its sediments were polluted by organic and metallic contaminants during the 1950–1980 period [48]. Twenty-one sites were sampled in Lake Saint-Louis (LSL), which receives waters from the Ottawa River in the north shore and from the St. Lawrence River in the south shore. In this fluvial lake, the waters and sediments of the north shore are enriched with organic carbon while those of the south-shore were polluted by metals, particularly mercury from industrial point sources in the 1960–1970 period [51–53]. Fifteen sites were sampled in Lake Saint-Pierre (LSP), the largest fluvial lake of the St. Lawrence River, 100 km downstream of Montreal. This large shallow lake is divided by the deep waterway, separating the north and south water masses. Three quarters of the lake area are covered with emergent and submerged vegetation, forming large wetlands inhabited by macroinvertebrates [16,17]. On the south shore, plumes from two rivers draining dairy farms and farmlands are point sources of nutrient and pesticide pollution, inducing the proliferation of benthic cyanobacteria [54]. Finally, thirteen sites were sampled in the Montreal Harbour area (MH), and downstream towards the Montreal municipality wastewater discharge plumes. The sediments in this area are the most heavily altered by organic and inorganic pollutants, and physical stress by regular maintenance dredging. All together, these sites represent the common habitats encountered in the fluvial section of the St. Lawrence River and cover a wide range of environmental conditions across a gradient of disturbances.

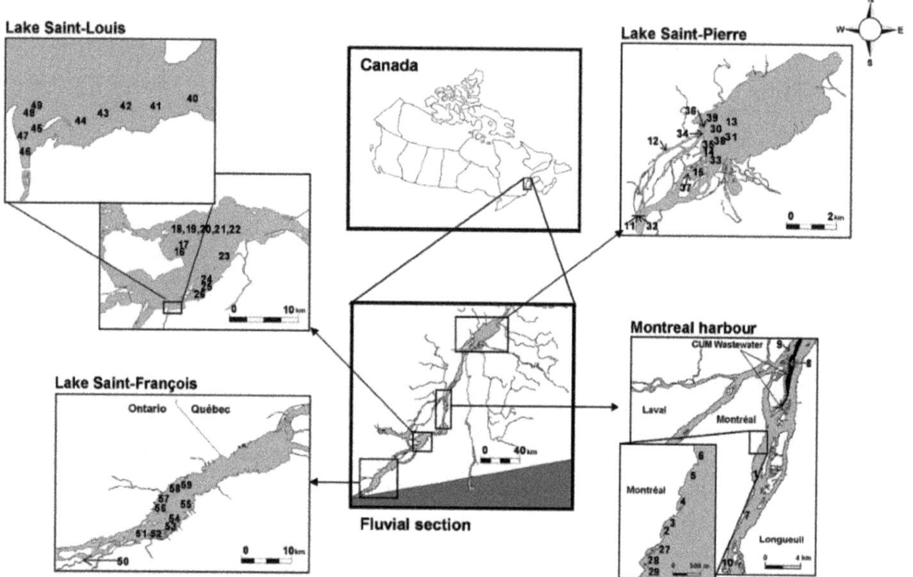


**Figure 1.** Sites sampled on four habitat zones across the fluvial section of the St. Lawrence River, including the three fluvial lakes and the Montreal Harbour area (from Figure 1, in [23,49]).

### 2.2. Sediment Sampling and Analysis

Twenty to twenty-five litres of surface sediments were collected using a Shipek grab sampler (400 cm$^2$) at each site and placed in clear polyethylene bags. In the field, sediment samples were kept on ice in containers for 24–30 h and thereafter stored at 4 °C in a cold chamber at the laboratory [23]. Up to 2 days after sampling, each sediment sample was manually homogenized and sieved through a 2 mm mesh size to retain coarse material prior to chemical analysis. The sediment analyses followed standard Quality Assurance and Control (QA/QC) protocols (see Table S1, Supplementary Materials for specific methods and detection limits in [49,50]). A wide range of sediment characteristics such as pH, total and dissolved organic carbon contents (TOC, DOC), particle grain size (% of sand, silt, clay, gravel) as well as the concentrations of nutrients, metalloids, metals and organic chemicals were analyzed (see Table S1, Supplementary Materials). Based on the sediment quality criteria established in the province of Quebec (Canada) for the remediation framework of contamination sites, the sediment in each site was classified into three quality classes (Table 1) according to the highest classification observed among all contaminants covered by the criteria [55]. Class 1 represents better sediments quality with all contaminants below the level of probable effect (PEL). Class 2 represents intermediate sediments quality with at least one contaminant between PEL and frequent effect levels (FEL). Class 3 represents lower sediments quality with at least one contaminant at a concentration higher than FEL. Class 3 sites were also divided into A and B categories according to whether the sites were contaminated and also under physical stress and degraded habitat (A: most of the Montreal Harbour sites) or only with historical contamination (1960–1980), without recent physical anthropogenic stress and better habitat condition (B: some of the sites in fluvial lakes). Sediment Class 1 includes 2 LSF, 9 LSL, 12 LSP and 2 MH Montreal wastewater sites. Sediment Class 2 includes 5 LSF, 4 LSL, 3 LSP and 2 MH sites. Sediment Class 3 includes 3 LSF, 8 LSL, 9 MH and no LSP site (Table 1, see Figure 1 for location sites). Spatial patterns in sediment quality classes reflect the gradient of disturbances. LSP sediments were generally of higher quality (Class 1 and Class 2). LSF sediments were also mostly of higher quality (Class 1 and Class 2) with only 3 sites in Class 3. LSL sediments were distributed in all classes, mostly in Class 1 and Class 3. LSL sites of Class 3 (considered for remediation) were in the south shore near a historical point source (River St. Louis) of mercury contamination [51]. Most of the MH sediments were of poor quality (Class 3) except for 2 sites in Class 2. In the 2 sites in the Montreal wastewater plume (PM8-PM9), sediments were of good quality (Class 1).

**Table 1.** Application of the sediment remediation framework and number of sites for each sediment quality class in the habitat zones (Lake Saint-François (LSF), Lake Saint-Louis (LSL), Lake Saint-Pierre (LSP) and Montréal Harbour (MH)).

| Sediment Quality Class | Remediation of Contaminated Site Application Framework [55] | LSF | LSL | LSP | MH |
|---|---|---|---|---|---|
| 1 | • All contaminant concentration < PEL (Probable Effect Level)<br>• Adverse biological effects may be anticipated<br>• Level of contamination alone does not justify site remediation | n = 2<br>LSF51 LSL53 | n = 9<br>LSL16 LSL17<br>LSL18 LSL19<br>LSL20 LSL22<br>LSL23 LSL24<br>LSL26 | n = 12<br>LSP12 LSP13<br>LSP14 LSP15<br>LSP30 LSP31<br>LSP33 LSP34<br>LSP35 LSP37<br>LSP38 LSP39 | n = 2<br>PM8 PM9 |
| 2 | • At least one contaminant concentration between PEL < concentration < FEL (Frequent Effect Level.)<br>• Identify the sources and take action If applicable eliminate inputs of contaminants Environmental studies may be necessary<br>• Assess the ecotoxicological risk<br>• Determine the remediation requirements | n = 5<br>LSF52 LSF56<br>LSF57 LSF58<br>LSF59 | n = 4<br>LSL21 LSL41<br>LSL43 LSL46 | n = 3<br>LSP11 LSP32<br>LSP36 | n = 2<br>PM2 PM5 |
| 3 | • At least one contaminant concentration > FEL<br>• Sediment contamination is considered a serious problem<br>• Identify the sources and take action If applicable eliminate inputs of contaminants<br>• Site remediation is desirable<br>• Biological assessments should be carried out to determine the feasibility of a remediation process, set the priorities for action and identify the environmental gains. | n = 3<br>LSF50 LSF54<br>LSF55 | n = 8<br>LSL25 LSL40<br>LSL42 LSL44<br>LSL45 LSL47<br>LSL48 LSL49 | n = 0 | n = 9<br>PM1 PM3<br>PM4 PM6<br>PM7 PM10<br>PM27 PM28<br>PM29 |

### 2.3. Macroinvertebrate Sampling and Analysis

Triplicate samples of macroinvertebrates were collected from the sediments using the Shipek grab (400 cm$^2$), placed in a polyethylene bag and preserved directly in the field with 10% formaldehyde solution. A Rose Bengal solution was added in each sample to stain macroinvertebrates and reduce sorting time. Back at the laboratory, the samples were rinsed with tap water on a 500 μm mesh size sieve, and the retained macroinvertebrates were transferred to 70% alcohol for subsequent identification [23]. Macroinvertebrates were counted and sorted in each replicate by a private company (SAB Laboratories Inc.); quality control was performed on 10% of the samples by a taxonomist expert (see [23,49] for more details). Variability in the number of taxa among replicated samples was low with less than 8% of new taxa sorted after analysing all triplicate samples compared to a single sample [49]. Macroinvertebrates were identified at the family and genus levels (except Nematoda) using several identification keys [56–58]. When the abundances of the dominant groups (Oligochaeta, Chironomidae and Gastropoda) were extremely high, a minimum of 100 individuals for these groups were randomly collected and identified. A complete list of the macroinvertebrate taxa recorded in the St. Lawrence River and used for this study is presented in Table S2 (Supplementary Materials). Dominance of macroinvertebrate taxonomical groups at each site served to establish spatial patterns in community composition among the habitat zones of the fluvial section, to calculate indices and metrics, and to determine the relationships with sediment characteristics and contamination.

*2.4. Metrics Selection, Scoring, and Statistical Analysis*

In the first step of the selection procedure (see Figure S1, Supplementary Materials), we made an inventory of all available macroinvertebrate indices and metrics based on taxon richness and diversity, abundance, and tolerance of taxa, multimetric biotic indices, and functional traits. This first selection gathered a total of 264 indices and metrics (see Table S3 in Supplementary Materials). Among these 264 indices and metrics, there were 21 usual indices of diversity and similarity, 72 metrics on richness, 93 indices and metrics on taxa abundances, 35 biotic indices, and 43 functional traits as recently proposed by Desrosiers et al. [49]. All indices and metrics that could not be calculated due to a lack of data (e.g., lack of data on taxa tolerance) or that were simply considered irrelevant to the current bioassessment in the St. Lawrence River (e.g., those used in small rivers and streams, or in foreign countries) were eliminated (33 indices and metrics in italics in Table S3). Finally, 231 indices and metrics were retained to test their relevance for assessing changes in macroinvertebrate community structure among the habitat zones of the fluvial section and among classes of sediment quality.

At the second step of the selection procedure (see Figure S1), we scored the 231 indices and metrics according to their potential to differentiate habitat zones (fluvial lakes and Montreal Harbour) and sediment quality classes by performing one-way ANOVAs (analysis of variance) by using a general linear model (GLM) with JMP (version 8.0.1, SAS Institute Inc Cary, NC, USA). Data distribution normality and variance equality were verified. In a small proportion, when the results did not meet these statistical requirements, we used a nonparametric test (Wilcoxon and Kruskal–Wallis). When analyses of variance were significant, differences among habitat zones or sediment classes were tested using Tukey–Kramer test for multiple comparisons. None of the indices and metrics without significant differences among habitat zones and classes of sediment quality were retained. Other indices and metrics showing significant differences among habitat zones (highlighted in blue in Table S3) or among sediment quality classes (highlighted in orange in Table S3), or among both habitat zones and sediment classes (highlighted in green in Table S3) were retained for further analysis. Since many of the indices and metrics could be redundant, a Spearman correlation table analysis among selected indices and metrics was performed to complement the analysis of variance to sustain our choices. This second procedure of selection based on analysis of variance and correlation analysis has enabled us to select 157 indices and metrics (in bold in Table S3) among the 231 ones retained after the first selection procedure.

In the third step of the selection procedure (see Figure S1), we applied principal component analysis (PCA) using CANOCO 4.0 [59,60] to determine the relevance of indices and metrics according to their collinearity and their potential to differentiate sampling sites in habitat zones. Collinearities among macroinvertebrate indices and metrics were determined by the angles between vectors ranging from 0° (maximum positive covariance) to 180° (maximum negative covariance), an angle of 90° indicating a lack of covariance [61]. A first PCA was carried out using the 157 indices and metrics retained after the second selection procedure. A stepwise sorting was performed by removing all indices and metrics with a low contribution in PCA ordination, i.e., close to the center or inside the circle of equilibrium contribution. Only those indices or metrics with a projection vector longer than the radius of the equilibrium contribution circle on the first two PCA axes were interpreted as the most suitable and relevant considering their potential to differentiate the sites representing a gradient of disturbance. The final selection was performed using PCA analyses with all sampling sites first and then with only the most representative sites (without extreme sites in LSL) to select the most parsimonious number of indices and metrics. After this third selection procedure, only 14 indices and metrics were retained (see Section 3).

In the fourth step of the selection procedure (see Figure S1) to establish a model linking the selected macroinvertebrate indices and metrics to sediment quality, we performed redundancy analyses (RDAs) using Monte Carlo unrestricted 999 permutation tests [61]. In the final RDA models, only the variables presenting significant relationships (prob. < 0.05) after stepwise selection were kept. The significance of the first three axes of the RDA was tested using the 'marginal' testing method using CANOCO 4.0 [59,60]. Finally, a regression tree methodology was applied using JMP software to develop predictive models determining contaminant thresholds which partitioned sites into homogeneous groups for each of the 14 final indices and metrics. This technique employs Euclidean distances to summarize between-site differences in community composition along changes in sediment contamination. Tree algorithms split the macroinvertebrate data set (assign macroinvertebrate data to groups) hierarchically (groups are then divided into subgroups) based on the ability of the contamination variables to predict changes in macroinvertebrate composition. This method is particularly useful to detect partitions (abrupt changes) in indices and metrics along a disturbance gradient in relation to significant thresholds in sediment contaminants [62,63].

## 3. Results

### 3.1. Typology of Macroinvertebrate Communities in the Fluvial Section

The sediments of the fluvial section of the St. Lawrence River supported abundant and diverse macroinvertebrate communities composed of fourteen taxonomic groups (Figure 2). Overall, the taxa belonged to 45 families and 109 genera (see Table S2 for the full list of taxa, Supplementary Materials). Macroinvertebrate composition varies among habitat zones and sites. In Lake Saint-Francois, macroinvertebrates were mainly composed of arthropods (Insecta), molluscs (Gastropoda), and crustaceans (Malacostraca). Community composition differed from upstream to downstream and between the north and south shores. Nematoda were found in greater abundance on the south shore than on the north shore while the Oligochaeta were relatively more abundant downstream than upstream, and inversely for the Malacostraca. In Lake Saint-Louis, community composition and dominance patterns were more variable from site to site than in Lake Saint-François. The Nematoda were more common on the north shore, the Gastropoda and Insecta in the bay of the island, and the Oligochaeta, Malacostraca and Bivalvia on the south shore. The Oligochaeta were more frequent downstream and the Nematoda upstream. In Lake Saint-Pierre, macroinvertebrate communities were also relatively diverse. Oligochaetes, Nematodes, Insects, and Bivalves were the most predominant groups. Community composition varied among the north and south shores, and along the longitudinal gradient. On both the north and south shores, communities were dominated by worms (Oligochaeta, Nematoda). However, Insecta and Bivalvia were more common in the north shore and at the upstream sites, while worms were more common at downstream sites. The Montreal Harbour supported the most disturbed community with a low diversity and a predominance of Oligochaeta associated to Insecta (mainly Diptera Chironomidae) and Bivalvia. Despite a large variability in macroinvertebrate composition among sites and habitat zones, the typology suggests a gradient of disturbance.

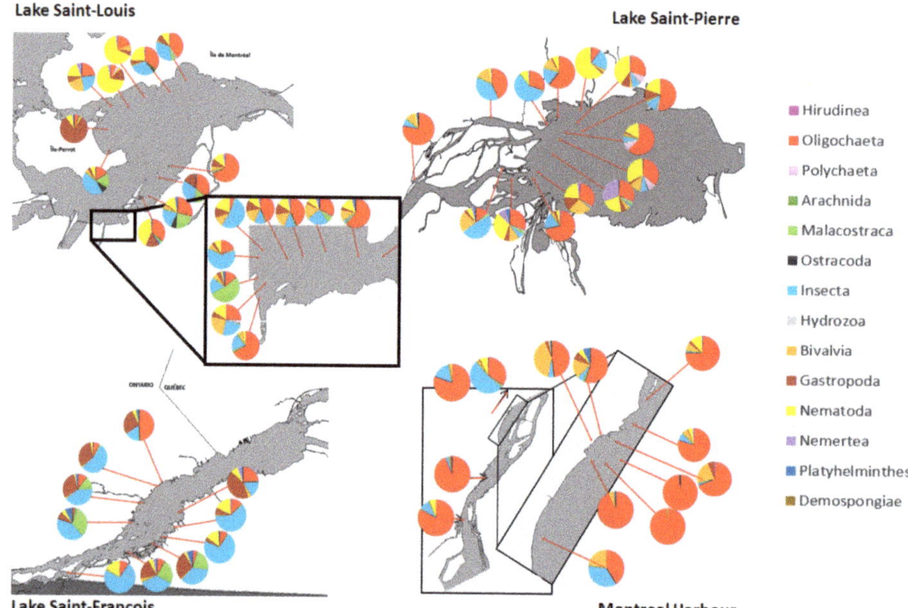



**Figure 2.** Relative abundance of macroinvertebrate communities in three fluvial lakes and the Montreal harbour along the fluvial section of the St. Lawrence River.

## 3.2. Selection of Macroinvertebrate Indices and Metrics

Selection procedures based on ANOVA and correlation analyses allowed us to select 157 indices and metrics (in bold in Table S3). These included 20 indices of diversity and similarity, 49 metrics of taxa richness, 36 metrics of taxa abundance and trophic guilds, 13 biotic indices based on taxa tolerance, and 39 functional traits (Table 2). Most of the selected indices and metrics (142) showed significant differences among habitat zones (highlighted in blue in Table S3). In contrast, only 5 metrics based on taxa abundance showed significant variation among sediment quality classes (ANEM, AHIR, ADIP, AHYD, AGOLD: highlighted in orange in Table S3), and only 3 metrics based on number of tolerant taxa and the dominance of scrapers, as well as 7 functional traits showed significant variation among both habitat zones and sediment quality classes (highlighted in green in Table S3).

The PCA analysis based on the 157 indices and metrics allowed us to eliminate additional 61 indices and metrics showing no significant contribution to spatial patterns in macroinvertebrate communities and sampling sites (Figure S2, Supplementary Materials). All fluvial lakes sites were grouped together in the center of the ordination plan, except for one extreme site in Lake Saint-Louis located in the lower right quadrant. The most impaired sites of Montreal Harbour were dissociated from those of the fluvial lakes in the upper right quadrant.

**Table 2.** List of the 157 metrics and indices retained after the first and second selection procedures. See Table S3 for full names and corresponding abbreviations. Taxa with significant difference among habitat zones (normal font); taxa with significant difference among sediment quality classes (italic font); taxa with significant differences among both habitat zones and sediment quality classes (underline font).

| | Indices and Metrics |
|---|---|
| Diversity indices | DSW, Hmax, ESW, DS, ES, M, BP, DH, DM, IG, PIE, TU, SIMF, SIMG, SHANF, SHANG, MARGF, MARGG, EVENF, EVENG |
| Metrics on taxa richness | NBFA, NBFOD, NBFEP, NBFTR, NBFDI, EPT, ETO, NBGE, NBGOD, NBGEP, NBGTR, NBGDI, NBGCH, NBTGaG, NBTGaF, NBTBiG, NBTBiF, NBTHiG, NBTHiF, NBTCrG, NBTCrF, NBTEG, NBTEF, NBTOdG, NBTOdF, NBTTrG, NBTTrF, NBTDiG, NBTDiF, NBTOtG, NBTOtF, NBTETG, NBTETF, NBTET/G, NBTET/F, NBTO/F, NBtEF, NBTOBF, NBTaxF, NBTODF, NBTODG, NBTEGA, NBTOLDG, NBTOD/G, NBTEG, NBTCOBG, NBFamG, NBGen, Ita |
| Metrics on taxa abundances | AOL, PFD, P2FD, P5FD, PCH, POL, PTR, PDI, PAM, PGA, PBI, PIN, %collector, %scraper, %predator, Mach%, NoIns%, %NemG, %GasG, %PolG, %OliG, %CruG, %TriG, %DipG, %HydG, %ETG, %ET/OLG, %AETG, %GOLDG, *ANEM*, APOL, AOLI, *AHIR*, *ADIP*, *AHYD*, *AGOLD* |
| Multi-metric biotic indices | IBGN, GFI, BMWP, ASPT, BBIF, BBIG, HAIF, HAIG, r/KF, r/KG, RETIF, RETIG, ICI |
| Metrics on functional traits [49] | Volt3, Life2, Sta3, Diss1, Diss3, Hab1, Hab2, Hab4, Att1, Att2, Att3, Att4, Form2, Form3, Arm1, Arm2, Arm3, Size2, Size4, Size5, Resp1, Resp2, Rep1, Rep2, Ovo2, Ovo3, Trop1, Trop4, Trop5, Foo1, Foo2, Foo4, Foo7, Vc1, Sub2, Sub3, Sub4, Sub5, Sub6 |

Finally, the stepwise RDA procedure comparing pairs of indices and metrics based on six criteria allowed us to eliminate another 143 indices and metrics and to retain only 14 metrics and indices as the most selective and parsimonious choices (Table 3). We selected the indices and metrics which were recognized as (1) relevant and easily to apply for bioassessment in large rivers, (2) having the potential to distinguish macroinvertebrate communities among habitat zones and/or sediment quality classes based on ANOVA analyses, (3) having the highest contributions in PCA ordination, and (4) showing significant relationships with sediment characteristics and contamination based on correlation analysis, RDA and regression tree analysis. We also gave priority to indices and metrics calculated at the genus level since our previous studies have shown a higher explanatory power at this taxon level [23,49] (5) and based on abundance and tolerance (6). Selective choices were made among indices and metrics that were collinear or had similar ecological principles. For instance, the metrics Ita and AOL were collinear ($\rho$ Spearman 0.95, $p = 0.9925$) (see also vector projections in Figure S2, Supplementary Materials). Thus, we retained only the metric AOL based on the abundance of Oligochaeta which was associated with the most impaired sites of the Montreal Harbour as shown with ANOVA and PCA analyses (Table 4, Figure S2). For example, we eliminated the metric Ital that was developed for small Italian rivers and judged inappropriate for a large and complex river such as the St. Lawrence River. The comparative selection of pairs of indices and metrics is detailed and presented in Table S4 (Supplementary Materials).

**Table 3.** List of taxa, metrics, and indices (with abbreviations) at the family (F) or genera (G) taxonomic levels, with equations, retained after all selection procedures.

| Taxa | Indices and Metrics | Abbreviations | G/F | Equations |
|---|---|---|---|---|
| Macroinvertebrates | Diversity of Menhinick | *DM* | F | $DM_n = \frac{S}{\sqrt{N}}$ |
| Macroinvertebrates | Diversity Shannon-Wiener-Index | *SHANG* | G | $SHANG = \sum\left(p_i \log_e p_i\right)$ |
| Oligochaeta | Taxonomic group (abundance) Oligochaeta | *AOL* | G | $AOL = \left(\frac{\sum \text{Oligochaeta}}{\text{Total effective}}\right)$ |
| Nematoda | Taxonomic group (abundance)—Nematoda | *ANEM* | G | $ANEM = \left(\frac{\sum \text{Nematoda}}{\text{Total effective}}\right)$ |
| Hirudinea | Taxonomic group (abundance)—Hirudinea | *AHIR* | G | $AHIR = \left(\frac{\sum \text{Hirudinea}}{\text{Total effective}}\right)$ |
| Gastropoda, Oligochaeta, Diptera | Taxonomic group (abundance)—GOLD | *AGOLD* | G | $AGOLD = \left(\frac{\sum \text{Taxons(Gastropoda+Oligochaeta+Diptera)}}{\text{Total effective}}\right)$ |
| Oligochaeta | Taxonomic group [%]—Oligochaeta [%] | *%Oli* | G | $\%Oli = \left(\frac{\sum \text{Oligochatea}}{\text{Total effective}}\right) \times 100$ |
| Diptera | Taxonomic group [%]—Diptera [%] | *%Dip* | G | $\%Dip = \left(\frac{\sum \text{Diptera}}{\text{Total effective}}\right) \times 100$ |
| Gastropoda, Oligochaeta, Diptera | Taxonomic group [%]—GOLD | *%GOLD* | G | $\%GOLD = \left(\frac{\sum \text{Taxa(Gastropoda+Oligochaeta+Diptera)}}{\text{Total effective}}\right) \times 100$ |
| Mollusca, Chironomidae, Achaeta | %abundance Mollusca, Achaeta, Chironomidae | *%Mach* | G | $\%Mach = \left(\frac{\sum \text{Taxa(Mollusca+Achaeta+Chironimidae)}}{\text{Total effective}}\right) \times 100$ |
| Non-insect | %abundance Non-insect relative | *NoIns%* | G | $\%Noinsc = \left(\frac{\sum \text{Taxons} \neq \text{insecta}}{\text{Total effective}}\right) \times 100$ |
| Macroinvertebrates | 5% first dominant families | *P5FD* | F | $P5FD = \left(\frac{\text{Effectif 5 dominant families}}{\text{Total effective}}\right) \times 100$ |
| Ephemeroptera, Trichoptera, Coleoptera, Odonata, Bivalvia | Taxonomic group (number of taxa)—ETCOB | *NBTCOBG* | G | $NBTCOBG =$ <br> $\sum \text{Taxa(Ephemeroptera} + \text{Trichoptera} + \text{Coleoptera} + \text{Odonata} + \text{Bivalvia)}$ |
| Ephemeroptera | Taxonomic group (number of taxa)— E-Taxa | *NBTETG* | G | $NBTETG = \text{Number of Ephemeroptera taxa}$ |

**Table 4.** Results of ANOVA analyses (classification and probabilities) for the 14 final metrics and indices considering variations among habitat zones (LSF-LSL-LSP-MH) or sediment quality classes based on the remediation framework (1-2-3) and separating sites with contamination and physical stress due to harbour activities (3A) and sites with only historical sediment contamination (3B). Non-significant tests (grey).

| Indices Metrics | Habitat Zones | | Sediment Quality Classes | | | |
|---|---|---|---|---|---|---|
| | LSF-LSL-LSP-MH | Prob. | 1-2-3 | Prob. | 1-2-3A-3B | Prob. |
| Diversity | | | | | | |
| DM | A-A-A-B | 0.0004 | A-A-A | 0.3832 | B-AB-C-A | <0.0001 |
| SHANG | A-A-A-B | 0.0003 | A-A-A | 0.8649 | B-B-C-A | <0.0001 |
| Taxa richness | | | | | | |
| NBTETG | A-A-B-B | <0.0001 | A-A-A | 0.2897 | BC-AB-C-A | <0.0001 |
| NBTCOBG | A-A-A-B | <0.0001 | A-A-A | 0.4273 | B-AB-C-A | <0.0001 |
| Taxa abundance | | | | | | |
| AOL | B-AB-AB-A | 0.0085 | A-A-A | 0.4617 | B-B-A-B | 0.0075 |
| %OliG | C-BC-B-A | <0.0001 | A-A-A | 0.1796 | B-B-A-B | <0.0001 |
| %DipG | A-B-B-B | 0.0005 | A-A-A | 0.9226 | AB-AB-B-A | 0.0176 |
| P5FD | BC-C-B-A | <0.0001 | A-A-A | 0.8449 | B-BC-A-C | <0.0001 |
| ANEM | A-A-A-A | 0.0846 | A-B-AB | 0.0181 | A-B-B-AB | 0.0380 |
| AHIR | A-A-A-A | 0.2213 | A-A-A | 0.6814 | A-A-A-A | 0.5122 |
| AGOLD | A-A-A-A | 0.4945 | A-A-A | 0.5070 | A-A-A-A | 0.5290 |
| %Match | A-B-B-B | <0.0001 | A-A-A | 0.8731 | AB-AB-B-A | 0.0176 |
| %NoInsc | B-A-A-A | 0.0003 | A-A-A | 0.9340 | AB-AB-A-B | 0.0058 |
| %GOLDG | AB-B-B-A | 0.0006 | A-A-A | 0.1202 | B-AB-A-B | 0.0025 |

The 14 indices and metrics selected were the most relevant for distinguishing habitat zones in the fluvial section (Figure 3). Increased diversity indices (SHANG, DM), greater richness in Ephemeroptera, Trichoptera, Coleoptera, Odonata and Bivalvia taxa (NBTETG, NBTCOBG), and higher abundances of Mollusca and Diptera Chironomidae (%Mach, %DipG, %GOLD, AGOLD) were associated primarily with Lake Saint-François sites (upper right and left quadrants), and opposed to certain sites of the Lake Saint-Pierre, Lake Saint-Louis and of the Montreal Harbour (lower left quadrant). In contrast, a higher percentage and abundance of non-insects (%NoIns) and worms such as Oligochaeta and Nematoda (ANEM, AOL, %OliG) were associated with certain impaired sites of the Lake Saint-Pierre, Lake Saint-Louis, and the Montreal Harbour (lower right quadrant).

Some of the final metrics were collinear (Figure 3) and based on similar ecological principles. For instance, the metrics %GOLD and AGOLD are redundant as well as the metrics %DipG and %Mach. Thus, for a more parsimonious selection, it may be appropriate to retain only those metrics with the highest contribution in the PCA ordination of sites such as AGOLD and %Match. Therefore, the ANEM, AHIR, SHANG, and NBTCOBG metrics could also be eliminated as they have the lowest contributions. The ANOVA analyses indicated which indices and metrics have the highest potential to discriminate habitat zones and sediment quality classes (Table 4) and complement the selection procedures.

Concerning habitat zones, diversity indices (DM, SHANG) distinguished only the sites of the fluvial lakes from the most impaired sites of the Montreal Harbour. Among metrics based on taxa richness, NBTCOBG also segregated the Montreal Harbour from the fluvial lake sites, while NBTETG segregated the LSF and LSL sites from the LSP and MH sites. The metrics %OliG and P5FD had a better potential than AOL, which distinguished only the two extreme habitat zones (LSF from MH). The metric %OliG segregated the MH impaired sites on one hand and the LSF sites on the other hand, but did not differentiate LSL and LSF sites, and LSP and LSL sites. The metrics %DipG, %Match and %NoIns segregated LSF sites from the sites of the two other fluvial lakes and the Montreal Harbour. The metrics ANEM, AHIR, and AGOLD failed to distinguish habitat zones, and the metric %GOLDG did not separate the less impaired LSF sites from the most impaired MH sites.

When considering the sediment quality classes, 12 final indices and metrics (except AHIR and AGOLD) presented a potential for distinguishing sediment quality classes when considering both categories of the class 3 (Table 4). Macroinvertebrate indices and metrics with the greatest potential to distinguish sediment quality classes were the diversity indices (DN, SHANG), the number of taxa at the genus level for the Ephemeroptera, and Trichoptera (NBTETG) as well as the total number of taxa of Trichoptera, Coleoptera, Bivalvia and Gastropoda (NBTCOBG). Most of them clearly differentiated the less impaired sites (Class 1) from the most impaired sites (Class 3A and 3B), indicating improved sediment quality. In contrast, the metrics based on the abundance and dominance of Oligochaeta (AOL, %OliG) discriminated the most contaminated sites (Class 3A), indicating lower sediment quality. The other metrics based on the abundance or percentage of tolerant taxa (%DipG, P5FD, ANEM, %Match, %NoIns, %GOLDG) presented the lowest potential to distinguish sediment quality classes. Although functional traits metrics performed slightly better than usual taxonomical metrics [49], they did not emerge as relevant in this sorting exercise. In addition, the database of traits available for macroinvertebrates of the St. Lawrence River is still incomplete. Consequently, trait approach is not currently used for bioassessment monitoring due to difficulty of their application in ecotoxicological risk assessment.

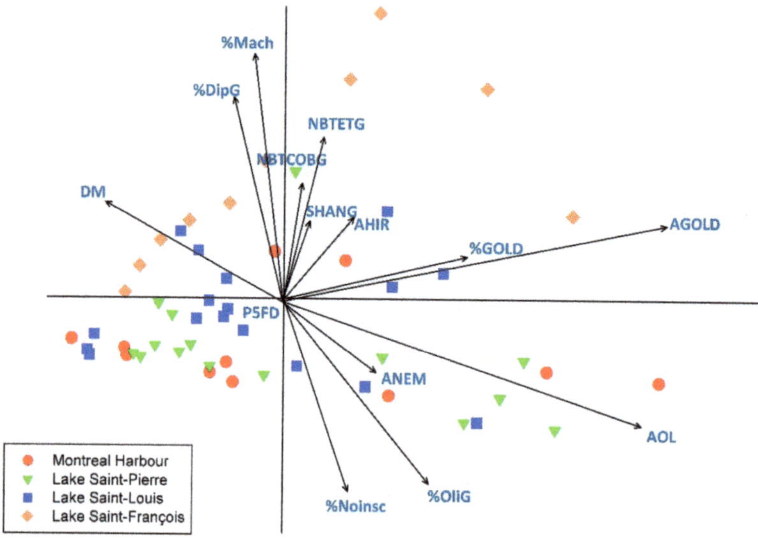


**Figure 3.** Principal component analysis showing the projection of the vectors of the 14 final indices and metrics and the ordination of the sampling sites of each habitat zone of the fluvial section (without 2 outliers of LSL sites).

### 3.3. Relationships between Macroinvertebrates Metrics and Indices and Sediment Characteristics and Contamination

To assess how the selected indices and metrics were related to sediment characteristics and contamination, we performed RDA analyses (Figure 4) and supplemented them with correlation analyses (Table S5). Only the results of RDA analysis without outlier sites are presented to provide a more comprehensive illustration of the relationships with sediment variables and the spatial distribution of sampling sites (see Figure S3, for the results with the outlier sites). In general, most of the indices and metrics showed higher significant correlations with sediments characteristics than contamination (Table S5).

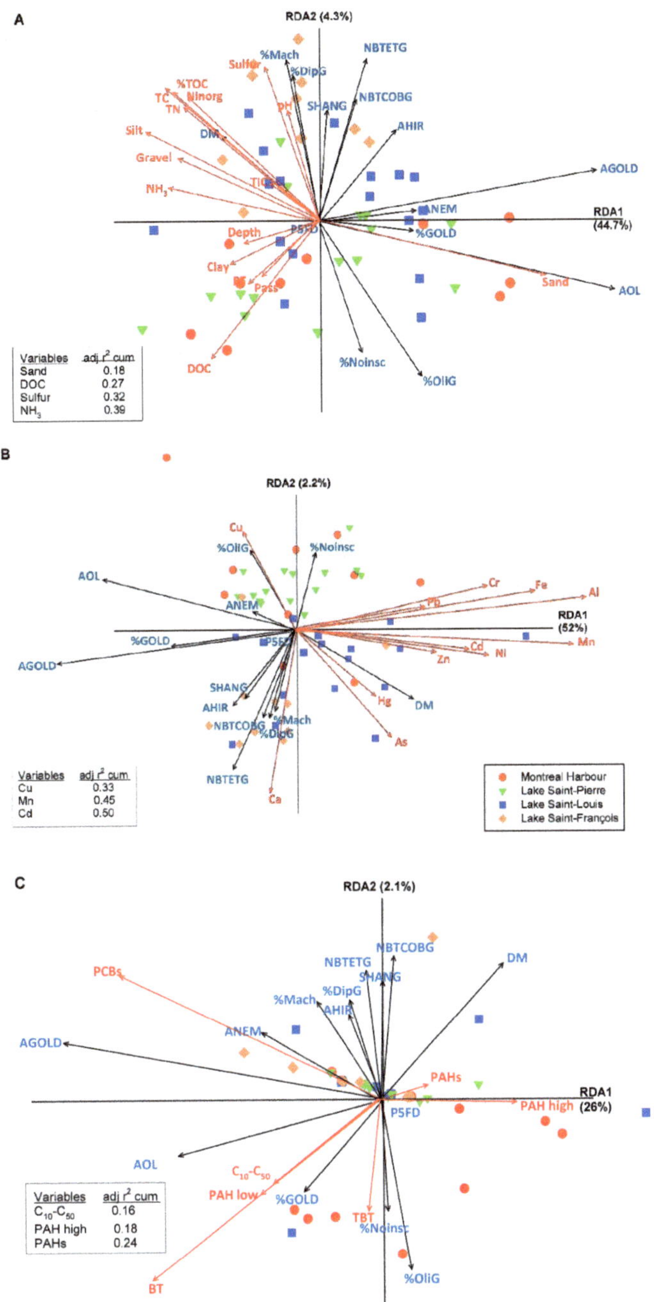

**Figure 4.** Redundancy analyses (RDA) based of the 14 selected indices and metrics and sediment characteristics and contaminants in 53 sampling sites of the three fluvial lakes and the Montreal harbour along the fluvial section of the St. Lawrence River. (**A**) sediment grain size composition and chemistry, (**B**) inorganic contaminants, (**C**) organic contaminants. Results without 6 outlier sites (LSL16, LSL17, LSL22, LSL26, PM28, PM6).

Overall, the RDA model relating indices and metrics to sediment characteristics explained 50% of the total variation in the macroinvertebrate community. We could dissociate two groups of metrics and indices depending on sediment composition, depth, nutrients and inorganic and organic carbon (Figure 4A). The significant explanatory variables were sand and $NH_3$ on axis 1, DOC and sulfur on axis 2 that explained 39% of the variance. Globally, along the first axis, metrics based on the abundance of tolerant taxa (AGOLD, AOL, %GOLD, ANEM) were associated with nutrient-poor sandy sediments (Sand, low $NH_3$) in shallow sites; other similar metrics (%NoIns, %OliG) were also associated with nutrient-poor sandy sediments (TN, Ninorg) and organic carbon (DOC, TOC, %TOC, TC). Most of these metrics had higher scores in the Montreal Harbour, Lake Saint-Louis, and Lake Saint-Pierre sites, which were considered to be the most impaired. On the second axis, diversity metrics (DM, SHANG), and metrics based on the relative abundance of ubiquist and sensitive taxa (% Match, %DipG) had higher scores in sediments composed of silt and gravel, with higher pH, sulfur, organic carbon (TOC, %TOC) and nitrogen (TN, Ninorg) levels. Taxa richness metrics (NBTETG, NBTCOBG) had higher scores in shallow sediments poor in nutrients (TP, Pass) and dissolved organic carbon (DOC). Most of these metrics had higher scores in the sites of Lake Saint-Louis and Lake Saint-François, which were considered as the less impaired.

Overall, the RDA model relating indices and metrics to sediment inorganic contamination explained 55.7% of the total variation in macroinvertebrate community. Given the trends observed with inorganic contaminants (Figure 4B), the sediments most polluted with metals and metalloids were found in the Montreal Harbour and the Lake Saint-Louis (see also Figure S3B). The significant explanatory variables were Mn and Cd on axis 1 and Cu on axis 2, which explained 50% of the variance. Diversity indices (DM, SHANG) and metrics based on richness or abundance of intolerant taxa (NBTETG, NBTCOBG, %Match, %DipG) were associated with sediments rich in calcium (Ca), mercury (Hg) and arsenic (As) but less contaminated in metals (Cd, Cu, Pb, Zn), most of which were found in the Lake Saint-François sites. The metrics based on tolerant taxa (%OliG) were associated with sediments rich in Cu at the Lake Saint-Pierre sites and other metals mainly at Montreal Harbour sites (relationships are better seen in Figure S3B). Spearman correlation analysis indicated significant positive or negative relationships between indices and metrics and inorganic contaminants (Al, As, Cd, Cr, Cu, Fe, Hg, Ni, Pb, Zn; Table S5).

Overall, the RDA model relating indices and metrics to sediment organic contaminants explained 28.1% of the total variation in the macroinvertebrate community. Given the trends observed for organic contaminants (Figure 4C), the sediments of the Montreal Harbour were the most contaminated by hydrocarbons and butyltins (PAHs, PAH High, PAH low, $C_{10}$-$C_{50}$ Petroleum Hydrocarbons, BT, TBT). The significant explanatory variables were $C_{10}$-$C_{50}$ petroleum hydrocarbon, PAH with high molecular weight (HAP high) and total PAH (PAHs). Here again, two types of metrics were opposed (better seen in Figure S3C). Metrics based on diversity indices and sensitive taxa (DM, SHANG, NBTETG, NBTCOBG, %Match, %DipG) were associated with the less contaminated sites in Lakes Saint-François, Saint-Louis, and Saint-Pierre. On the other hand, metrics based on tolerant taxa (AOL, %OliG, %GOLD, %NoIns) were associated with the most oil- or butyltin-polluted sediments in the Montreal Harbour.

*3.4. Responses of Metrics and Indices to Contaminant Thresholds*

Cascading homogenous grouping thresholds were determined for the 14 selected indices and metrics using inorganic and organic contaminants in regression tree models. For each index and metric, the estimated thresholds were compared to (i) the criteria established to assess the sediment quality in a remediation context [55], and (ii) natural concentrations in sediments during preindustrial period < 1950 and postglacial clays [55,64] (See Table S6A,B). A total of 10 over 14 indices and metrics showed robust tree regression models ($r^2$ > 60%). There were divided into two groups: (1) diversity (DM, SHANG), richness (NBTETG, NBTCOBG) and dominance (P5FD) metrics based on ubiquitous and sensitive taxa changed mainly with inorganic contaminants (Pb, Zn, Hg, Ca) rather than with organic contaminants (PCBS, PAHs), (2) metrics based on the abundance of tolerant taxa (%GOLD,

%OliG, %Match, %Noinsc) were more related to organic contaminants (PCBS, PAHs) than inorganic contaminants (Cu, As, Cd). An example of a regression tree model for each type of indices and metrics is presented in Figure 5. The other models are presented in the Figure S4, Supplementary Materials.

According to the criteria established for the sediment quality [55], most of the thresholds determined by the regression trees were below Probable Effect Level (PEL) or Frequent Effect Level (FEL), but still above concentrations in preindustrial sediments except for As (6.6 mg/kg), and in postglacial clays except in some case for Ni, Cr and Cu (75, 150 and 54 mg/kg respectively) or ambient levels except for As (2–7 mg/kg) and Cr (52–93 mg/kg) [55,64].

For the DM model ($r^2$ = 0.65), Pb concentration below the PEL was the first node discriminating 51 sites with a concentration below 61 mg/kg with DM values of 1.35 ± 0.40, followed by calcium dividing these sites into two blocks with 23 sites below and 28 sites above 19,000 mg/kg of calcium. In sites with lower Ca concentration, PCBs were taken into account for the classification of the sites, most sites with PCBs content below 0.0423 mg/kg were discriminated a second time by calcium (11,000 mg/kg) and PAHs (1.66 mg/kg). In sites with Ca concentration above 19,000 mg/kg, contamination by Cd, Hg and Cu below the PEL threshold completed the site classification.

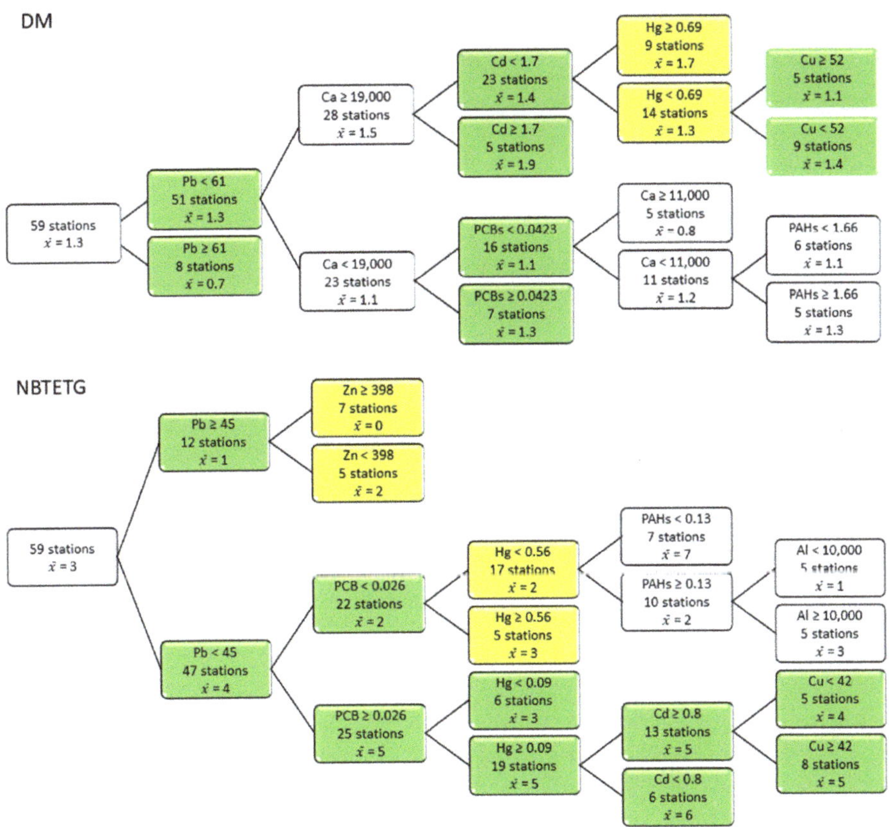

**Figure 5.** *Cont.*

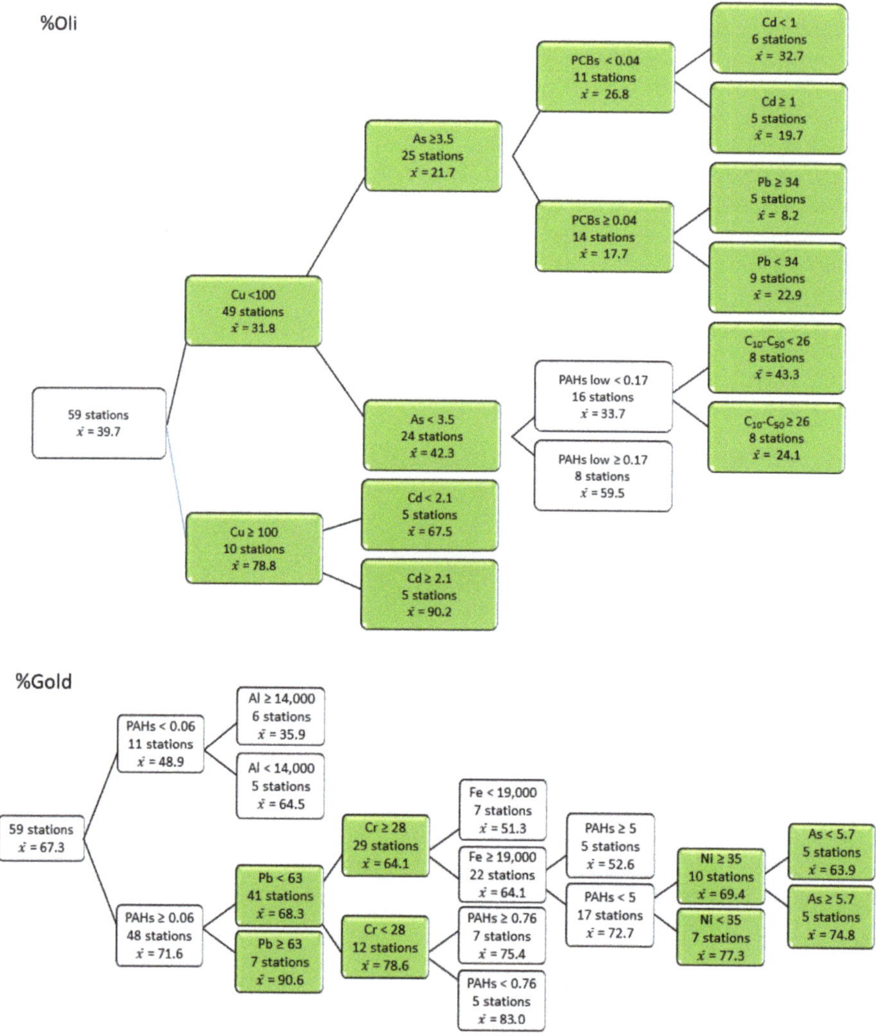

**Figure 5.** Regression tree models for a diversity index (DM), and metrics based on taxa richness (NBTETG) and taxa abundances (%OliG, %GOLD). Green represents the thresholds below the PEL, yellow between the PEL and the FEL and white are the substances without sediment quality criteria.

For the NBTETG model, a Pb concentration below the PEL was again the first node discriminating 47 sites below and 12 sites above 45 mg/kg. The 12 sites with higher Pb concentration were divided by a Zn concentration between PEL and FEL (398 mg/kg). The group of 7 sites with higher Pb and Zn contamination had the lowest NBTETG value (0.14 ± 0.38). The 47 sites below the threshold of 45 mg/kg was followed by a separation of sites in two blocks based on PCBs contamination threshold (0.026 mg/kg) below the PEL. At the sites with PCBs concentrations below this threshold (22 sites), Hg, PAHs and Al explained the classification of sites. At sites with higher PCBs concentrations (25 sites), Hg, Cd and Cu explained the classification of the sites.

For the %OliG model, a Cu concentration below the PEL was the first node discriminating 49 sites with higher Cu concentrations (100 mg/kg). For sites with higher Cu concentrations, the Cd concentration was the second node (2.1 mg/kg) separating the sites in two equal numbers (5 sites each). The group of 5 sites with higher Cu and Cd contamination had the highest %OliG value (90.20 ± 13.10%). At the sites with low Cu content, As (3.5 mg/kg) at the lower than pre-industrial concentration was the second node dividing the sites into two equal blocks. At sites with higher As content, PCBs was the third node, and metal contamination by Cd or Pb was the last node. At sites with low As, organic contaminants (PAHs, $C_{10}$-$C_{50}$ Petroleum hydrocarbons) were the last nodes.

For the %GOLD model, organic contamination by PAHs was the first node separating 48 sites with higher PAHs concentrations. Pb concentration (63 mg/kg) below the PEL segregated another 41 stations with lower Pb concentration and 7 stations with higher concentration. The 7 sites with higher PAHs and Pb concentration had the highest %GOLD value (90.61 ± 9.71%). The 41 sites with low Pb concentration the second node was divided by Cr (28 mg/kg). The sites with concentration below this threshold (12 stations) was divided by PAHs and the sites with concentration above this threshold (29 stations) were divided by Fe (22 stations), PAHs contamination (17 stations), followed by metal contaminants (Ni, As).

Overall, models for the diversity indices and metrics based on richness (DM, SHANG) and the number and abundance of ubiquitous or intolerant taxa (NBTETG, NBTCOBG, P5FD, %Match) were more related to changes in inorganic Pb and Zn contamination for the first nodes, and organic contamination by PCBs for the second nodes. Metrics based on relative abundance of tolerant taxa (%OliG, %NoIns, %DipG) were also related to changes in inorganic Cu contamination. Metrics based on the abundance of tolerant taxa (%GOLD, AOL, AHIR, ANEM, AGOLD) were more related to organic contamination by PAHs contamination for the first nodes and inorganic contamination by Pb, Cu, and Mn for the second nodes.

## 4. Discussion

### 4.1. The St. Lawrence River Case Study

The St. Lawrence River does not fit to the River Continuum Model [65], since it is not oriented along a longitudinal gradient (upstream to downstream). It forms a complex hydrological network made up of different water masses and a mosaic of habitats along its transversal dimension (from the uplands to the channel). This corresponds to the complex river model developed by Thorp and collaborators [66,67]. In accordance with the Serial Discontinuity Model [68], fluvial lakes and river confluences disrupt the St. Lawrence River continuum by creating discontinuities and riparian zones of sediment deposition invaded by macrophytes [18]. Compared to small rivers with specific pollution sources where changes in macroinvertebrate indices and metrics can easily be detected and monitored between sites upstream and downstream of the pollution point source [40], large rivers are impacted by multiple anthropogenic stressors and diffuse pollution sources that can interact in multiple ways with environmental conditions [24,46].

In the St. Lawrence River, the absence of defined stress gradients and the wide variation in benthic communities among habitats are important challenges for the development of bioassessment programs. Previous studies in the Lake Saint-Pierre have compared macroinvertebrate assemblages at reference sites and impacted tributary plume sites downstream of two rivers draining agricultural lands [17]. Macroinvertebrate communities in sediments were dominated by endobenthic fauna such as Oligochaeta and Sphaeridae. Taxon richness and composition differed between the reference fluvial sites and the impacted tributary plume sites, reflecting the variation in sediment metal contamination. Oligochaeta and Diptera Chironomidae were characteristic of the impacted sites, while the Gastropoda Valvatidae and the Amphipoda Gammaridae and Asellidae were more abundant in the reference sites. Other studies conducted at the scale of the fluvial sector of the St. Lawrence River analysed spatial patterns of variation of macroinvertebrate communities as a function of taxonomic composition [23] or functional traits [49]. Four macroinvertebrate assemblages were found distributed in the different zones of the fluvial continuum in relation to habitat characteristics and sediment contamination [23].

Four groups of sites were also defined using the functional trait-based approach, which performed better than the taxonomic approach in differentiating sites along the sediment contamination gradient [49]. However, none of these studies evaluated the performance of various macroinvertebrates-based indices and metrics to assess changes in sediments characteristics and contamination. Our case study in the fluvial continuum of the St. Lawrence River is a new complement to previous studies conducted in American [15,24] and Russian [46] large rivers.

### 4.2. Rational for the Selection and Inclusion of Indices and Metrics

Only a limited number (14) of indices and metrics from an initial selection of 157 were deemed relevant to assess sediment quality in the fluvial section of the St. Lawrence River (Tables 2 and 3). Our final score was composed of 2 diversity indices (DM, SHANG), 2 metrics based on the number of sensitive taxa (NBTETG, NBTCOBG), and 10 metrics based on the abundance and dominance of sensitive (%Match, %NoInsc, %DipG) or tolerant (%OliG, %GOLD, AOL, AHIR, ANEM, AGOLD) taxa. This score is similar to this of 9 of the 97 benthic metrics used in the bioassessment of six large tributaries of the upper Mississippi and Ohio rivers in Midwest United States [15]. Their final index was composed of metrics based on the number of taxa and dominance of sensitive (Diptera, EPT and Coleoptera taxa richness) or tolerant (tolerant taxa richness, % of Oligochaete and leech taxa) organisms, and metrics based on trophic groups (% of collector-filterer, burrower, and facultative taxa, predator taxa richness). Our final score also corresponds to the 12 indices and metrics included in the PANEL multimetric index developed for the Ohio River, considered responsive to water quality disturbance and ecologically relevant [40].

The diversity indices included in this study (DM, SHANG) are based on the premise that stable and healthy benthic communities at reference sites are taxonomically richer than those at sites impacted by anthropogenic disturbances (e.g., acid drainage, nutrient enrichment, sediment contamination) [40]. Total taxon richness has also been reported as a reliable index in different Austrian river types [38]. Macroinvertebrate diversity indices are considered not robust metrics for biological assessment [29]. Indeed, they are more related to ecological factors such as productivity and habitat heterogeneity than to ecotoxicological and disturbance factors [29], except in cases of extreme pollution. This is also the case of the St. Lawrence River, where the selected diversity indices (DM, SHANG) based on genus level distinguished only the most disturbed sites of the Montreal Harbour from the fluvial sites (Table 4). However, they have an interesting potential for distinguishing sediment quality classes (Table 4).

Metrics based on the numbers of sensitive taxon at the genus level of insects (Ephemeroptera, Trichoptera: NBTETG), and on the richness of facultative taxa (Trichoptera, Coleoptera, Odonata and Bivalvia: NBTCOBG) were effective in differentiating habitat zones and sediment quality classes (Table 4). They show consistently high discriminatory power for less impaired sites and better water and sediment quality [69]. These sensitive organisms are highly intolerant to pollution and their species richness and abundance decrease as anthropogenic disturbances increase in rivers [40], lakes [12] and streams [14,38]. Ephemeroptera larvae are the most sensitive and the first group to disappear in the presence of anthropogenic disturbances, while Trichoptera larvae are considered moderately tolerant. They have the best BMWP scores (6–10) for predicting the deleterious effects of pesticides, metals and organic contaminants on benthic communities in small rivers [20]. In the St. Lawrence River, these metrics showed a wide variation, but allowed us to differentiate between the most impacted sites in the Montreal Harbour and the less impaired sites in the fluvial lakes (mainly LSF and LSL), and the sediment quality classes (Table 4). The potential of these metrics reflects their relevance as indicator taxa of good ecological status in previous studies based on taxonomic composition and functional traits of macroinvertebrates. In the St. Lawrence River, Ephemeroptera (Ephemeridae) and Trichoptera (Hydroptilidae and Leptoceridae) were indicator taxa of the reference sites and less impaired LSF, LSL, and LSP sites [23]. They belong to functional groups composed of univoltine insect larvae with collector and shredder feeding modes, sensitive to organic pollution [49]. Hydroptilidae is mainly a family of large rivers and constitutes one of the most sensitive taxa of Trichoptera [40].

Coleoptera (Elmidae), Odonata and Bivalvia were also sensitive taxa found in less impaired sites in the St. Lawrence River.

Metrics based on the dominance of tolerant taxa (%OliG, %DipG, %Match, %Noinsc, %GOLDG, %P5FD) were also useful for differentiating habitat zones and sediment quality classes (Table 4). They indicate many conditions in moderately to highly disturbed sites, and lower sediment quality. Diptera, especially Chironomidae, show tolerances to many factors but increase under disturbed conditions, as does the percentage of non-insects in rivers [40] and streams [38]. Worms such as Oligochaeta and Dipteran Chironomidae have the worst BMWP scores (1–2) and can support high levels of pollution by pesticides, metals and organic contaminants in small rivers [20]. Although this is true when just looking at a family level of Chironomidae, this is not true anymore when you look at the species level, as the structure and species assemblage of the family Chironomidae changes substantially in lakes [70] and also in the St. Lawrence river [23,49]. Gastropoda and Amphipoda share intermediate BMWP scores (3–6) and are relatively tolerant pollution stress. In the St. Lawrence, the metric based on the 5 most dominant groups (%P5FD) is composed of these tolerant taxa (Oligochaeta, Diptera, Gastropoda) and has been associated with disturbed conditions. The metric based on the percentage of highly tolerant taxa such as the Oligochaeta (%OliG) was the most associated with impaired sites in the Montreal Harbour (Table 4). However, metrics based on the abundance of tolerant taxa (AOL, ANEM, AHIR, AGOLD) had less or no potential to differentiate between habitat zones and sediment quality classes.

### 4.3. Comparative Responses to Habitat Characteristics and Sediment Contamination

Distinguishing anthropogenic and natural influences and effects on ecosystems is a fundamental problem in environmental sciences [46]. This is problematic in bioassessment of large rivers with macroinvertebrates, where most of the currently applied indices and metrics depend on natural environmental factors. The St. Lawrence River case study is another highlight of the importance of habitat characteristics above sediment contamination. On the other hand, in extreme conditions of pollution or disturbance in the Montreal Harbor, identifying of indices and metrics specific to sediment contamination in large rivers is challenging due to the high natural variability and diversity of factors affecting macroinvertebrate communities [71]. Understanding the relationships between different bioassessment indices and metrics and natural environmental factors is necessary to assess their lack of correlation.

### 4.4. Bioassessment in a Large and Complex River: Limit and Pertinence

The hydrology, water quality and riparian habitats of the St. Lawrence River have been substantially altered in recent decades [72,73]. Water level fluctuations related to climatic conditions, can strongly affect emergent and submerged aquatic vegetation and macroinvertebrate communities on plants and sediments [16,18,74]. For the bioassessment of the ecological status of large and complex rivers as the St. Lawrence, special attention is needed to assess the impacts of: (1) hydromorphological alterations; (2) activities such as dredging; and (3) harbor and industrial development on biological communities, especially those that affect river continuity and riparian cover [2]. One of the challenges in large rivers is to distinguish the effects of natural ecological factors and anthropogenic stressors. This is to support ecological risk assessment and management in river systems and to assess the complex interactions between multiple stressors.

### 4.5. Conclusions and Recommendation for Bioassessment Programs

Macroinvertebrates are commonly used for the bioassessment of rivers subject to anthropogenic disturbance. This approach generally requires comparison with reference conditions. Unfortunately, it is not always easy to predict the response of macroinvertebrate communities to environmental changes either anthropogenic or simply related to natural habitat variations. This is particularly the case in large rivers with a mosaic of highly diverse habitats. Indeed, in this study, the typology of macroinvertebrate communities varies according to fluvial lakes, water bodies and sediment quality. Fourteen indices and metrics were shown to be most effective in differentiating between sites and quality classes, and it is these parameters that are most likely to be used in bioassessment for the St. Lawrence River or other large and complex river in northern temperate regions. However, the indices and metrics remain strongly explained by habitat characteristics, such as sediment grain size or the presence of nutrients. There is also an influence of metals and, to a lesser extent, organic contaminants such as petroleum hydrocarbons. The predictive power of indices and metrics is also higher than what has been observed in our previous studies of community structure using taxonomy or functional traits [23,49]. This makes the 14 selected indices and metrics promising bioassessment tools while being easier to use, interpret and explain in an environmental assessment context.

Projects in the St. Lawrence River will continue, and more data will be needed to establish management thresholds. On the other hand, analysis of the results obtained from the regression trees highlights changes in the structure of the macroinvertebrate community below the FEL or even PEL, allowing for the detection of more subtle and early effects. The use of %Oligochaeta seems a very promising variable to detect the presence of metals, but also to distinguish the combined effect with that of hydrocarbons.

**Supplementary Materials:** The following are available online at http://www.mdpi.com/2073-4441/12/12/3335/s1, Table S1. Minimum and maximum values of the sediment characteristics in the 59 sampling stations: pH, total and dissolved organic carbon (TOC, DOC), and concentrations of nutrients, metals, metalloid and organic chemicals (mg/kg DW). <DL = under detection limit, min and max, Table S2. Macroinvertebrate taxa at genus level recorded in the fluvial section of the St. Lawrence River and used for this case study: Occurrences based on number of sites and frequency (%), Table S3. List of the 264 macroinvertebrate metrics and indices (with their abbreviations) colligated from the literature including 181 usual and 83 new (with asterisks) metrics and indices. (A) Usual and news metrics and indices. (B) Functional traits indices. The 157 metrics and indices retained after ANOVAs and correlations analyses are in bold. Metrics highlighted in blue showed significant differences among habitat zones; metrics highlighted in orange showed significant differences among sediment quality classes; metrics highlighted in green showed significant differences among habitat zones and sediment quality classes (ANOVAs, $p < 0.01$), Table S4. Step by step pair comparisons (index 1 vs index 2) of the 157 macroinvertebrate metrics and indices (with their abbreviations) based on several criteria: (1) their pertinence in bioassessment of large rivers (Per), (2) their potential for discriminating habitat zones and class of sediment quality (ANOVA), (3) their correlations with sediment variables (Cor), (4) their contribution in the PCA ordination (PCA+), (5) their easy identification at the genus level (Gen), and (6) as quantitative metrics based on taxa abundance (AB). Taxa in bold are those selected step by step for each pair of metrics or indices. The metrics based on functional traits were not retained because of their difficulty to be applied by managers in current bioassessment, Table S5. Correlations (ρ de Spearman) among macroinvertebrate metrics and indices retained after the selection procedure and the characteristics of habitat and sediment (* $p < 0.05$; ** $p < 0.01$; *** $p < 0.0001$), Table S6. A- Criteria used for the evaluation of sediments quality, Table S6. B - Criteria used for the evaluation of sediments quality, Figure S1. Macroinvertebrates indices and metrics selection procedure framework, Figure S2. Principal component analysis (PCA) based of the 157 metrics and indices retained after the first selection using ANOVAs, and of sampling sites in three fluvial lakes and the Montreal harbour along the sections of the St. Lawrence River, Figure S3. Redundancy analyses (RDA) based of the 14 selected metrics and indices and sediment characteristics and contaminants in all sampling sites of the three fluvial lakes and the Montreal harbour along the fluvial section of the St. Lawrence River. (A) grain size, composition, and chemistry, (B) inorganic contaminants, (C) organic contaminants, Figure S4. Regression tree models. Green represents the thresholds below the PEL, yellow between the PEL and the FEL and white are the substances without sediment quality criteria.

**Author Contributions:** M.D. at MELCC and B.P.-A. at the Université de Montréal managed the project. They conceived the framework, supervised the data collated and analysis of the Master student, and wrote the English version of the paper. C.S. collated all data on macroinvertebrate indices and metrics from previous studies, ran the statistical analysis and wrote the French report. All authors have read and agreed to the published version of the manuscript.

**Funding:** This study is a component of a collaborative program included in the St. Lawrence Plan for Sustainable Development, and realised with the active participation of Environment and Climate Change Canada (Environmental Protection Operations Division and Science and Technology Branch), the Ministère de l'Environnement et de la Lutte contre les changements climatiques (Centre d'expertise en analyse environnementale du Québec; Direction des évaluations environnementales des projets hydriques et industriels; Direction générale du suivi de l'état de l'environnement). C. Spilmont was supported by a scholarship grant from a Discovery funding of the National Science and Engineering Research Council (NSERC) to B.P.A.

**Acknowledgments:** We thank Gaëlle Triffault-Bouchet for the revision of the manuscript, and collaborative training of the Master student. Macroinvertebrates were identified by a private firm (Laboratoires SAB Inc.).

We also thank M. Arseneault, G. Brault, A. Lajeunesse and P. Turcotte, who participated to fieldwork and all persons connected to the laboratories of chemistry of the Centre d'expertise en analyse environnementale du Québec. We thank Louise Cloutier (Université de Montréal), for their help about taxonomic identification.

**Conflicts of Interest:** The authors declare no conflict of interest.

## References

1. Hering, D.; Feld, C.K.; Moog, O.; Ofenböck, T. Cook book for the development of a Multimetric index for biological condition of aquatic ecosystems: Experiences from the European AQEM and STAR projects and related initiatives. *Hydrobiologia* **2006**, *556*, 311–324. [CrossRef]

2. Reyjol, Y.; Argillier, C.; Bonne, W.; Borja, A.; Buijse, A.D.; Cardoso, A.C.; Daufresne, M.; Kernan, M.; Ferreira, M.T.; Poikane, S.; et al. Assessing the ecological status in the context of the European water framework directive (WFD): Where should we go now? *Sci. Total Environ.* **2014**, *497/498*, 332–344. [CrossRef] [PubMed]

3. Pinel-Alloul, B.; Méthot, G.; Borcard, D. *Évaluation de L'état des Données sur les Macroinvertébrés Benthiques du Fleuve Saint-Laurent et de ses Tributaires Pour une Application d'un Modèle de Suivi Environnemental avec L'approche des Conditions de Référence (CABIN)*; Centre Saint-Laurent, E.C., Ed.; Université de Montréal: Montréal, QC, Canada, 2004; p. 68.

4. U.S. Environmental Protection Agency. *Methods for Evaluating Wetland Conditions: Developing an Invertebrate Index for Biological Integrity of Wetlands Office of Water*; U.S.E.P.A.: Washington, DC, USA, 2002.

5. Chessman, B.C. Rapid assessment of rivers using macroinvertebrates: A procedure based on habitat-specific sampling, family-level identification, and a biotic index. *Aust. J. Ecol.* **1995**, *20*, 122–129. [CrossRef]

6. Chessman, B.C.; Royal, M.L. Bioassessment without reference sites: Use of environmental filters to predict natural assemblages of river macroinvertebrates. *J. North Am. Benthol. Soc.* **2004**, *23*, 599–615. [CrossRef]

7. Metzeling, L.; Chessman, B.; Hardwick, R.; Wong, V. Rapid assessment of rivers using macroinvertebrates: The role of experience, and comparisons with quantitative methods. *Hydrobiologia* **2003**, *510*, 39–52. [CrossRef]

8. Bonada, N.; Prat, N.; Resh, V.H.; Statzner, B. Developments in aquatic insect biomonitoring: A comparative analysis of recent approaches. *Annu. Rev. Entomol.* **2006**, *51*, 495–523. [CrossRef]

9. Resh, V.H.; Norris, R.H.; Barbour, M.T. Design and implementation of rapid assessment approaches for water resource monitoring using benthic macroinvertebrates. *Aust. J. Ecol.* **1995**, *20*, 108–121. [CrossRef]

10. Cayrou, J.; Compin, A.; Giani, N.; Céréghino, R. Associations spécifiques chez les macroinvertébrés benthiques et leur utilisation pour la typologie des cours d'eau. Cas du réseau hydrographique Adour-Garonne (France). *Ann. Limnol. Int. J. Limnol.* **2000**, *36*, 189–202. [CrossRef]

11. Tripole, S.; Vallania, E.A.; Corigliano, M.C. Benthic macroinvertebrate tolerance to water acidification in the Grande River sub-basin (San Luis, Argentina). *Limnetica* **2008**, *27*, 29–38.

12. Gabriels, W.; Lock, K.; De Pawn, N.; Goethals, P.L.M. Multimetrics macroinvertebrate Index Flanders (MMIF) for biological assessment of rivers and lakes in Flanders (Belgium). *Limnologica* **2010**, *40*, 199–207. [CrossRef]

13. Lewin, I.; Czerniawska-Kusza, I.; Lawniczak, A.E.; Jusik, S. Biological indices applied to benthic macroinvertebrates at reference conditions of mountain streams in two ecoregions (Poland, the Slovak Republic). *Hydrobiologia* **2013**, *709*, 183–200. [CrossRef]

14. Alemneh, T.; Ambelu, A.; Zaitchik, B.F.; Bahrndorff, S.; Mereta, S.T.; Pertoldi, C. Macroinvertebrate multi-metric index for Ethiopian highland streams. *Hydrobiologia* **2019**, *843*, 125–141. [CrossRef]

15. Blocksom, K.A.; Johnson, B.R. Development of a regional macroinvertebrate index for large river bioassessment. *Ecol. Indic.* **2009**, *9*, 313–328. [CrossRef]

16. Tessier, C.; Cattaneo, A.; Hudon, C.; Borcard, D. Invertebrate communities associated with metaphyton and emergent and submerged macrophytes in a large river. *Can. J. Fish. Aquat. Sci.* **2008**, *70*, 10–20. [CrossRef]

17. Tall, L.; Méthot, G.; Armellin, A.; Pinel-Alloul, B. Bioassessment of benthic macroinvertebrates in wetland habitats of Lake Saint-Pierre (St. Lawrence River). *J. Great Lakes Res.* **2008**, *34*, 599–614. [CrossRef]

18. Tall, L.; Armellin, A.; Pinel-Alloul, B.; Méthot, G.; Hudon, C. Effects of hydrological regime, landscape features, and environment on macroinvertebrates in St. Lawrence River wetlands. *Hydrobiologia* **2016**, *778*, 221–241. [CrossRef]

19. Armitage, P.D.; Moss, D.; Wright, J.T.; Furse, M.T. The performance of a new biological water quality score system based on macroinvertebrates over a wide range of unpolluted running water sites. *Water Res.* **1983**, *17*, 333–347. [CrossRef]

20. Armitage, P.D. The potential of RIPACS for predicting the effects of environmental change. Chapter 7. In *Assessing the Biological Quality of Freshwaters: RIPACS and Other Techniques*; Wright, J.F., Sutcliffe, D.W., Furse, M.T., Eds.; Freshwater Biological Association: Ambleside, UK, 2000; pp. 93–111.

21. Burt, A.J.; McKee, P.M.; Hart, D.R.; Kauss, P.B. Effects of pollution on benthic invertebrate communities of the St. Marys River. *Hydrobiologia* **1991**, *2019*, 63–81. [CrossRef]

22. Clements, W.H.; Cherry, D.S.; Van Hassel, J.H. Assessment of the impact of heavy metals on benthic communities at the Clinch River (Virginia): Evaluation of an index of community sensitivity. *Can. J. Fish. Aquat. Sci.* **1992**, *49*, 1686–1694. [CrossRef]

23. Masson, S.; Desrosiers, M.; Pinel-Alloul, B.; Martel, L. Relating macroinvertebrate community structure to environmental characteristics and sediment contamination at the scale of the St. Lawrence River. *Hydrobiologia* **2010**, *647*, 35–50. [CrossRef]

24. Angradi, T.R.; Pearson, M.S.; Bolgrein, D.W.; Jicha, T.M.; Taylor, D.L.; Hill, B.H. Multimetric macroinvertebrate indices for mid-continent US great rivers. *J. North Am. Benthol. Soc.* **2009**, *28*, 785–804. [CrossRef]

25. Reynoldson, T.B.; Norris, R.H.; Resh, V.H.; Day, K.E.; Rosenberg, D.M. The reference condition: A comparison of multimetric and multivariate approaches to assess water-quality impairment using benthic macroinvertebrates. *J. North Am. Benthol. Soc.* **1997**, *16*, 833–852. [CrossRef]

26. Bailey, R.C.; Kennedy, M.G.; Dervish, M.Z.; Taylor, R.M. Biological assessment of freshwater ecosystems using a reference condition approach: Comparing predicted and actual benthic invertebrate communities in Yukon streams. *Freshw. Biol.* **1998**, *39*, 765–774. [CrossRef]

27. Czerniawska-Kusza, I. Comparing modified biological monitoring working partly score system and several biological indices based on macroinvertebrates for water quality assessment. *Limnologica* **2005**, *35*, 169–176. [CrossRef]

28. Washington, H.G. Diversity, biotic and similarity indices: A review with special relevance to aquatic ecosystems. *Water Res.* **1984**, *18*, 653–694. [CrossRef]

29. Koperski, P. Diversity of freshwater macrobenthos and its use in biological assessment: A critical review of current applications. *Environ. Rev.* **2011**, *19*, 16–31. [CrossRef]

30. Buffagni, A.; Erba, S.; Cazzola, M.; Murray-Bligh, J.; Soszka, H.; Genoni, P. The STAR common metrics approch to the WFD intercalibration process: Full application for small, lowland rivers in three European countries. *Hydrobiologia* **2006**, *566*, 379–399. [CrossRef]

31. Reynoldson, T.B.; Rosenberg, D.M.; Resh, V.H. Comparison of models predicting invertebrate assemblages for biomonitoring in the Fraser River catchment, British Columbia. *Can. J. Fish. Aquat. Sci.* **2001**, *58*, 1395–1410. [CrossRef]

32. Odoutan, O.H.; Janssens de Bisthoven, L.; Abou, Y.; Eggermount, H. Biomonitoring of lakes using macroinvertebrates: Recommended indices and metrics for use in West Africa and developing countries. *Hydrobiologia* **2019**, *826*, 1–23. [CrossRef]

33. Oertli, B.; Indermuehle, N.; Angélibert, S.; Hinden, H.; Stoll, A. Macroinvertebrate assemblages in 25 high alpine ponds of the Swiss National Parks (Cirque of Macun) and relation to environmental variables. *Hydrobiologia* **2008**, *597*, 29–41. [CrossRef]

34. Walley, W.J.; Grbowitč, J.; Dzeroski, S. A reappraisal of saprobic values and indicator weights based on Slovenian river quality data. *Water Res.* **2001**, *35*, 4285–4292. [CrossRef]

35. Hilsenhoff, W.L. Rapid field assessment of organic pollution with a family-level biotic index. *J. North Am. Benthol. Soc.* **1988**, *7*, 65–68. [CrossRef]

36. Hilsenhoff, W.L. An improved biotic index for organic stream pollution. *Great Lakes Entomol.* **1987**, *20*, 31–39.

37. Pinel-Alloul, B.; Méthot, G.; Lapierre, L.; Willsie, A. Macroinvertebrate community as a biological indicator of ecological and toxicological factors in Lake Saint-François (Québec). *Environ. Pollut.* **1996**, *91*, 65–87. [CrossRef]

38. Ofenböck, T.; Moog, O.; Gerritsen, J.; Barbour, M. A stressor specific multimetric approach for monitoring running waters in Austria using benthic macro-invertebrates. *Hydrobiologia* **2004**, *516*, 251–268. [CrossRef]

39. Touron-Poncet, H.; Bernadet, C.; Compin, A.N.B.; Céréghino, R. Implementing the Water Framework Directive in overseas Europe: A multimetric macroinvertebrate index for river bioassessment in Caribbean islands. *Limnologica* **2014**, *47*, 34–43. [CrossRef]

40. Applegate, J.M.; Baumann, P.C.; Emery, E.B.; Wooten, M.S. First steps in developing a multimetric macroinvertebrate index for the Ohio River. *River Res. Appl.* **2007**, *23*, 683–697. [CrossRef]

41. Zamora-Muñoz, C.; Alba-Tercedor, J. Bioassessment of organically polluted Spanish rivers, using a biotic index and multivariable methods. *J. North Am. Benthol. Soc.* **1996**, *15*, 332–352. [CrossRef]

42. Barton, D.R.; Metcalfe-Smith, J.L. A comparison of sampling techniques and summary indices for assessment of water quality in the Yamaska River, Quebec, based on benthic macroinvertebrates. *Environ. Monit. Assess.* **1992**, *21*, 225–244. [CrossRef]

43. Karr, J.R.; Fausch, K.D.; Angermeier, P.L.; Yant, P.R.; Scholosser, U. *Assessing Biological Integrity in Running Waters: A Method and Its Rationale*; Illinois Natural History Survey Special Publication No 5: Champaign, IL, USA, 1986.

44. De Pauw, N.; Vanhooren, G. Method for biological quality assessment of watercourses in Belgium. *Hydrobiologia* **1983**, *100*, 153–168. [CrossRef]

45. Armanini, D.G.; Horrigan, N.; Monk, W.A.; Peters, D.L.; Baird, D.J. Development of a benthic macroinvertebrate flow sensitivity index for Canadian rivers. *River Res. Appl.* **2011**, *27*, 723–737. [CrossRef]

46. Schletterer, M.; Fureder, L.; Kusolev, V.V.; Beketov, M.A. Testing the coherence of several macroinvertebrates indices and environ mental factors in a large lowland river system (Volga River, Russia). *Ecol. Indic.* **2010**, *10*, 1083–1092. [CrossRef]

47. Collier, K.J.; Hamer, M.P.; Moore, S.C. Littoral and benthic macroinvertebrate community responses to contrasting stressors in a large New Zealand river. *N. Z. J. Mar. Freshw. Res.* **2014**, *48*, 560–576. [CrossRef]

48. Carignan, R.; Lorrain, S.; Lum, K. A 50-yr record of pollution by nutrients, trace metals, and organic chemicals in the St. Lawrence River. *Can. J. Fish. Aquat. Sci.* **1994**, *51*, 1088–1100. [CrossRef]

49. Desrosiers, M.; Usseglio-Polatera, P.; Archaimbault, V.; Larras, F.; Méthot, G. Pinel-Alloul, B. Assessing anthropogenic pressure in the St. Lawrence River using traits of benthic macroinvertebrates. *Sci. Total Environ.* **2019**, *649*, 233–246. [CrossRef]

50. Desrosiers, M.; Babut, M.P.; Pelletier, M.; Bélanger, C.; Thibodeau, S.; Martel, L. Efficiency of sediment quality guidelines for predicting toxicity: The case of the St. Lawrence River. *Integr. Environ. Assess. Manag.* **2010**, *6*, 225–239.

51. Désy, J.C.; Archambault, J.F.; Pinel-Alloul, B.; Hubert, J.; Campbell, P.G.C. Relationships between total mercury in sediments and methyl mercury in the freshwater gastropod prosobranch *Bithynia tentaculata* in the St. Lawrence River, Quebec. *Can. J. Fish. Aquat. Sci.* **2000**, *57*, 164–173.

52. Flessas, C.; Couillard, Y.; Pinel-alloul, B.; St-Cyr, L.; Campbell, P.G.C. Metal concentration in two freshwater gastropods (Mollusca) in the St. Lawrence River and relationships with environmental contamination. *Can. J. Fish. Aquat. Sci.* **2000**, *57*, 126–137. [CrossRef]

53. Amyot, M.; Pinel-Alloul, B.; Campbell, P.G.C. Abiotic and seasonal factors influencing trace metal levels (Cd, Cu, Ni, Pb, and Zn) in the freshwater amphipod *Gammarus fasciatus* in two fluvial lakes of the St. Lawrence River. *Can. J. Fish. Aquat. Sci.* **1994**, *51*, 2003–2016. [CrossRef]

54. Lévesque, D.; Cattaneo, A.; Hudon, C.; Gagnon, P. Predicting the risk of proliferation of the benthic cyanobacterium *Lyngbia wollei* in the St. Lawrence River. *Can. J. Fish. Aquat. Sci.* **2012**, *69*, 1585–1595. [CrossRef]

55. Environnement Canada, Ministère du Développement durable, de l'Environnement et des Parcs du Québec. *Criteria for the Assessment of Sediment Quality in Quebec and Application Frameworks: Prevention, Dredging and Remediation*; Plan Saint-Laurent pour un développement durable: Québec, QC, Canada, 2007; p. 39.

56. Klemm, D.J.; Lewis, P.A.; Fulk, F.; Lazorchak, J.M. *Macroinvertebrate Field and Laboratory Methods for Evaluating the Biological Integrity of Surface Waters*; U.S Environmental Protection Agency, Office of Research and Development of the Environmental Monitoring Systems Laboratory: Cincinnati, OH, USA, 1990; p. 206.

57. Merritt, R.W.; Cummins, K.W. *An Introduction to the Aquatic Insects of North America*, 3rd ed.; Kendall/Hunt Publishing Company: Dubuque, Iowa, 1996; p. 442.

58. Thorp, J.H.; Covich, A.P. *Ecology and Classification of North American Freshwater Invertebrates*, 2nd ed.; Academic Press: San Diego, CA, USA, 2001; p. 1038.

59. Ter Braak, C.J.F. Canonical correspondence analysis: A new eigenvector technique for multivariate direct gradient analysis. *Ecology* **1996**, *67*, 1167–1179. [CrossRef]

60. Ter Braak, C.J.F.; Smilauer, P. *CANOCO Reference Manual and User's Guide to Canoco for Windows: Software for Canonical Community Ordination (Version 4)*; Microcomputer Power: Ithaca, NY, USA, 1998.

61. Legendre, P.; Legendre, L. Numerical Ecology. In *Development in Environmental Modelling*, 3rd ed.; Elsevier Science BV: Amsterdam, The Netherlands, 2012; Volume 24, p. 1006.

62. De'ath, G. Multivariate regression trees: A new technique for modeling species-environment relationships. *Ecology* **2002**, *83*, 1105–1117.

63. Ouelette, M.-H.; Legendre, P.; Borcard, D. Cascade multivariate regression tree: A novel approach for modelling nested explanatory sets. *Methods Ecol. Evol.* **2012**, *3*, 234–244. [CrossRef]

64. Saulnier, I.; Gagnon, C. Background levels of metals in St. Lawrence River sediments: Implications for sediment quality criteria and environmental management. *Integr. Environ. Assess. Manag.* **2006**, *2*, 126–141. [CrossRef] [PubMed]

65. Vannote, R.L.; Minshall, G.W.; Cummins, K.W.; Sedell, J.R.; Cushing, C.E. The river continuum concept. *Can. J. Fish. Aquat. Sci.* **1980**, *37*, 130–137. [CrossRef]

66. Thorp, J.H.; Thoms, M.C.; Delong, M.D. The riverine ecosystem synthesis: Biocomplexity in river networks across space and time. *Res. Appl.* **2006**, *22*, 123–147. [CrossRef]

67. Thorp, J.H.; Thoms, M.C.; Delong, M.D. *The Riverine Ecosystem Synthesis: Towards Conceptual Cohesiveness in River Science*; Academic Press: London, UK, 2008; p. 232.

68. Stanford, J.A.; Ward, J.V. Revisiting the serial discontinuity model. River. *River Res. Appl.* **2001**, *17*, 303–310. [CrossRef]

69. Wallace, J.B.; Grubaugh, J.W.; Whiles, M.R. Biotic indices and stream ecosystem processes: Results from an experimental study. *Ecol. Appl.* **1996**, *6*, 140–151. [CrossRef]

70. Dorić, V.; Pozojević, I.; Vučković, N.; Ivković, M.; Mihaljević, Z. Lentic chironomid performance in species-based bioassessment proving: High-level taxonomy is not a dead end in monitoring. *Ecol. Indic.* **2020**, *107041*. in press.

71. Beketov, M.A.; Liess, M. An indicator for effects of organic toxicants on lotic invertebrate communities: Independence of confounding environmental factors over an extensive river continuum. *Environ. Pollut.* **2008**, *156*, 980–987. [CrossRef]

72. Hudon, C. Impact of water-level fluctuations on St. Lawrence River aquatic vegetation. *Can. J. Fish. Aquat. Sci.* **1997**, *54*, 2853–2865. [CrossRef]

73. Hudon, C.; Carignan, R. Cumulative impacts of hydrology and human activities on water quality in the St. Lawrence River (Lake Saint-Pierre, Québec, Canada). *Can. J. Fish. Aquat. Sci.* **2008**, *65*, 1165–1180. [CrossRef]

74. Hudon, C.; Gagnon, P.; Amyot, J.-P.; Létourneau, G.; Jean, M.; Plante, C.; Rioux, D.; Deschênes, M. Historical changes in herbaceous wetland distribution induced by hydrological conditionsin Lake Saint-Pierre (St. Lawrence River, Quebec, Canada). *Hydrobiologia* **2005**, *539*, 205–224. [CrossRef]

**Publisher's Note:** MDPI stays neutral with regard to jurisdictional claims in published maps and institutional affiliations.

*Article*

# Spatial and Temporal Patterns of Macroinvertebrate Assemblages in the River Po Catchment (Northern Italy)

**Riccardo Fornaroli** [1,*], **James C. White** [2], **Angela Boggero** [1] **and Alex Laini** [3]

1   CNR—Water Research Institute (IRSA), Corso Tonolli 50, 28922 Verbania, Italy; angela.boggero@irsa.cnr.it
2   River Restoration Centre, Cranfield University, Cranfield, Bedfordshire MK43 0AL, UK;
    J.C.White@cranfield.ac.uk
3   Department of Chemistry, Life Sciences and Environmental Sustainability, University of Parma,
    Parco Area delle Scienze 11/a, 43124 Parma, Italy; alex.laini@unipr.it
*   Correspondence: riccardofornaroli@gmail.com

Received: 1 July 2020; Accepted: 28 August 2020; Published: 31 August 2020

**Abstract:** In the last decade, large scale biomonitoring programs have been implemented to obtain a robust understanding of freshwater in the name of helping to inform and develop effective restoration and management plans. A comprehensive biomonitoring dataset on the macroinvertebrate assemblages inhabiting the rivers of the Po Valley (northern Italy), comprised a total of 6762 sampling events (period 2007–2018), was analyzed in this study in order to examine coarse spatial and temporal trends displayed by biotic communities. Our results showed that macroinvertebrate compositions and derived structural and functional metrics were controlled by multiple environmental drivers, including altitude and climate (large scale), as well as habitat characteristics (local scale). Altitude proved to be the primary geographic driver, likely due to its association with thermal and precipitation regimes, thus explaining its overriding influence on macroinvertebrate assemblages. Significant temporal variations were observed across the study period, but notably in 2017, the overall taxonomic richness and diversity increased at the expense of Ephemeroptera, Plectoptera and Trichoptera taxa during an unprecedented heatwave that occurred across southern Europe. The detail of this study dataset allowed for important environmental attributes (e.g., altitude, habitat characteristics) shaping biotic communities to be identified, along with ecologically vulnerable regions and time periods (e.g., extreme climatic events). Such research is required globally to help inform large-scale management and restoration efforts that are sustainable over long-term periods.

**Keywords:** bioassessment; temporal trend; altitude; climate; insects

---

## 1. Introduction

Assessing the ecological status of riverine ecosystems is fundamental to informing the conservation and management of freshwater ecosystems [1]. A large amount of data has been generated through biomonitoring efforts globally, including in the USA, Australia, South Africa and various nations within Europe [2]. Such data often span decadal time periods and cover broad spatial scales, making them ideal to uncover spatio-temporal effects of environmental drivers on biological communities. Examining data generated from biomonitoring programs is thus crucial to guide river management strategies and achieve ecological improvements through identifying primary environmental drivers shaping riverine ecosystems, as well as ecologically important regions and time periods. Moreover, such information can be useful in informing water managers and environmental regulators on how to enhance efficiency of the existing network of biomonitoring sites (e.g., biologically important locations).

Rivers serve many societal functions and are one of the most intensively human influenced ecosystems worldwide [3]. The benefits of water provision for various human purposes, including agriculture, energy production, drinking water supply, and the displacement of pollutant loads often impair riverine ecosystems and biodiversity [4], with potentially serious societal costs. Anthropogenic activities threaten riverine ecosystems through habitat loss and degradation [5] such as physical modification of riverine environments, deforestation of pristine wildernesses, pollution and introduction of non-native species [6–8]. Riverine environments, due to their societal and ecological importance, are integral to environmental policies such as the Water Framework Directive (WFD—[9]), and benthic invertebrate fauna (macroinvertebrates) are one of the predominant Biological Quality Elements [9] used for their ecological assessment. Macroinvertebrates are intermediately positioned in freshwater food webs, consuming basal resources like primary producers, organic matter and detritus, and serving as a food source for amphibians, birds, reptiles, fish and humans.

Both biodiversity assessments and biomonitoring programs utilizing macroinvertebrate assemblages routinely process and examine single metrics targeting different facets of community assemblages, such as taxonomic richness (e.g., [10]); taxonomic diversity (e.g., [11,12]); ecological response guild summaries (e.g., [13–15]); community abundance (e.g., [16]) and functional diversity (e.g., [17]). However, metrics characterizing different community properties respond differently to environmental stressors. Thus, single-metric approaches can miss important information about the assemblage structure and its response to environmental gradients [18], highlighting the importance of adopting multi-metric approaches to increase the likelihood of detecting ecological responses to different environmental conditions and stressors. In summary, variations in environmental conditions driven by natural and anthropogenic processes can mediate its influence on different structural and functional metrics. As such, other multivariate approaches (e.g., Non-Metric Multidimensional Scaling—NMDS) may thus be used along with metrics to better represent assemblage structure variations.

Despite the large amount of data collected on Italian river macroinvertebrates since the implementation of WFD programs, only recently has such data become available for researchers, policy makers and water managers (e.g., [19–21]). Fornaroli and coauthors [21] collected and homogenized data on the presence, distribution and abundances of macroinvertebrate taxa inhabiting the River Po catchment (northern Italy) in the last decade, providing the first checklist of macroinvertebrates occurring in this area. This data source represents a comprehensive spatial and temporal data set on the macroinvertebrate assemblages of the rivers of the Po Valley to help overcome the lack of analyses in the region and to help guide effective restoration and management efforts.

The objective of this study is to examine spatial and temporal trends of the macroinvertebrate assemblages inhabiting the River Po catchment. We tested the following hypotheses: (H1) macroinvertebrate assemblages show a geographic pattern that is primarily driven by altitude; (H2) there are differences among the assemblages inhabiting the pristine and anthropogenically altered study sites; (H3) there are families associated with the human alteration status; (H4) macroinvertebrate response metrics display significant temporal trends.

## 2. Materials and Methods

### 2.1. Study Area and Samples Selection

For this study we selected suitable data from a recently released dataset [21] collating the data collected by the Environmental Agencies using the Italian national standardized method [22] for the implementation of WFD activities. The underpinning data were collected between 2007 and 2018 and comprised a total of 6762 sampling events. All study rivers lie within the River Po catchment (North Italy), which covers 71,000 km$^2$ and crosses different Administrative Regions (Aosta Valley, Piedmont, Lombardy, Liguria, Veneto, Trentino, and Emilia-Romagna) including mostly the Subalpine area and the Po Plain, (Biogeographic region: Alpine, Continental and Mediterranean, [23]). Following the national standardised monitoring protocol [22], three samples were collected each year with a

Surber-net enclosing an area of 0.05 or 0.10 m$^2$, with a mesh size of 0.5 mm. Each sample consisted of ten replicates collected proportionally to reflect predominance of the most representative microhabitats (mix of substratum type, current velocity, presence/absence of macrophytes, etc.). The different sampling efforts (0.5 or 1 m$^2$—i.e., all replicates of each Surber-net summed) were standardized by reporting all the abundances to 1 m$^2$. The protocol includes sampling each site in at least two different seasons (the most related to the local life cycle of insects) [11]; consequently, the winter season is underrepresented in this dataset and excluded from the analyses.

To allow for the evaluation of temporal trends, only sites sampled for at least three years and with at least two sampled seasons annually were retained. Among the selected sites, none belong to the Emilia-Romagna administrative region because they do not fulfill the minimum requirements of replicates that we imposed for temporal trend analyses, as samples collected before 2013 are not present in the original database. The 270 remaining sites (Figure 1) had an average (median) of 11 samples, collected over an average period of 7 years. The considered rivers are natural, artificial (channel) or partially modified by different forms of infrastructure (e.g., dams, weirs) constructed along the rivers. For the selected samples sites, we extracted percentage representation of sampled microhabitats, classified in nine mineral substrate classes (e.g., cobble and gravel) and eight biological substrates (e.g., emerging macrophytes and coarse particulate organic matter) as included in the national standardized method [22], the presence of human impacts (i.e., each site was a priori classified as pristine or altered), the WGS-84 coordinates (as decimal degrees) and altitude (as m a.s.l.). The altitude of the sampling sites ranges from 29 to 2280 m a.s.l.

Identification of organisms was performed at different taxonomic rank levels by operators of the Environmental Agencies, thus, we homogenized the taxonomic level to the least common denominator (mostly family level) using the "biomonitoR" package [24] within R software [25].

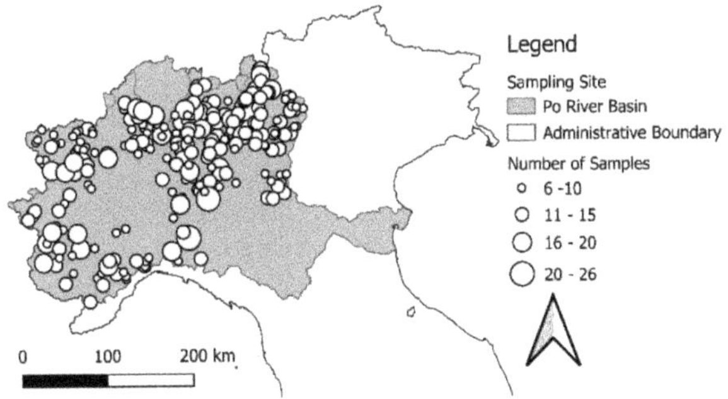


**Figure 1.** Map of the study area with the location of sampling sites.

### 2.2. GIS Analyses

We extracted several climatic variables from the WorldClim database (version 1.4) at the highest resolution (30 arc-s) using the function *getData* from the "raster" package [26] for each study site. These variables represented mean monthly and annual values of temperature and precipitation. For subsequent analyses, we considered 18 bioclimatic variables derived from the monthly temperature and rainfall values in order to generate more biologically meaningful variables. The bioclimatic variables represent annual trends (e.g., mean annual temperature, annual precipitation) seasonality (e.g., annual range in temperature and precipitation) and extreme or limiting environmental factors (e.g., temperature of the coldest and warmest month, and precipitation of the wet and dry quarters).

### 2.3. Macroinvertebrate Metrics

We extracted the relative density (i.e., log$_{10}$(individual per m$^2$ + 1) of three macroinvertebrate families (Plecoptera: Nemouridae, Ephemeroptera: Heptageniidae and Trichoptera: Limnephilidae) that were selected based on these families being: (i) identified as driving assemblage differentiation along altitudinal gradient; (ii) widely used as good environmental indicators for lotic ecosystems (e.g., [13,14]); and (iii) being the more abundant family within the respective order during the study period. We also evaluated three assemblage metrics commonly used in the bioassessment of riverine environments: (i) the richness of all families; (ii) the percentage representation of Ephemeroptera, Plecoptera and Trichoptera (%EPT) individuals; and (iii) the Shannon diversity index (H') [11]. Such metrics and indices represent different parameters of the benthic assemblages such as taxa abundance, sensitivity and diversity.

Based on the same list of macroinvertebrates, each family in the database was linked to its auto-ecological preferences downloaded from the freshwaterecology.info database [27,28]. Freshwaterecology.info is an online tool that unifies, standardizes and codifies more than 20,000 European freshwater organisms and their biological properties and ecological preferences: through this database, we obtained traits of more than 140 macroinvertebrate families. The nomenclature of functional traits is reported herein by their "grouping features" and "traits" (see [29]). Grouping features represent a functional trait category (e.g., "substrate preference" and "food"), while traits signify modalities residing within these (e.g., substrate preference—"gravel", "macrophytes"; food—"fine detritus"," dead animal"). A total of 21 grouping features comprising 113 traits were utilized from the functional traits database in subsequent analyses (Table S1). We derived functional traits from Tachet and coauthors [30] and, subsequently, aggregated them in two functional diversity indices.

Firstly, the Functional Richness (FRic) was derived, which represents the amount of functional space filled by the community [31]. FRic is defined by the trait extremes and thus reflects the potential maximum functional dissimilarity. FRic is calculated as the hypervolume enclosing the functional space filled by the community, the resulting FRic variable is standardized by its maximum, ranging from 0 to 1. Secondly, Rao's quadratic entropy [32] was used to estimate Functional Diversity (FDiv) because it has been considered more appropriate than other indices [33]. This index is estimated using abundance data and standardized by the maximum value to constrain the values within the range 0–1.

### 2.4. Data Analyses

We used non-Metric Multidimensional Scaling (NMDS), a gradient analysis approach based on a distance or dissimilarity matrix, to visualize differences in taxonomic structure of the macroinvertebrate assemblage [34]. NMDS is an iterative procedure and was performed using the function *metaMDS* from the "vegan" package [35]. For this, we used the Bray-Curtis dissimilarity distance, a non-Euclidean distance used to quantify the compositional dissimilarity between two different samples, and $\log_{10}(x + 1)$ transformed abundance in each sample. To examine the gradient of effect of the geographic variables, we fitted smooth surface splines using the *ordisurf* function from the "vegan" package. This procedure uses generalized additive models (GAMs) to overlay a smoothed response surface, which allows a more detailed interpretation than a simple linear vector [36].

To assess the correlation between macroinvertebrate assemblages and environmental variables, we ran variance partitioning [37,38] for the macroinvertebrate assemblage matrix and the metrics, using the *varpart* function in "vegan" while the significance of fractions was tested using the *dbrda* function. This method partitions the variation between the pure effects of each variable, or group of variables and the shared variance explained. First, we considered the three groups of variables (i.e., geographic, climatic or habitat) together, and afterwards we ran separate analyses for each group. The pure effect of each geographic variable (i.e., altitude, latitude or longitude) was considered, while for habitats the variables were divided in two groups (distribution in classes of mineral and biological substrates abundance) as well as for climatic variables (temperature and precipitation related variables).

Differences in assemblage composition among alteration classes were quantitatively explored, as well as seasonal controls, with the additive effects of "sampling season" and "alteration status" being tested within a permutational multivariate analysis of variance (PERMANOVA) via the *adonis* function within the "vegan" package. Indicator species analysis (IndVal.g; [39]) was employed to search for significant indicator species discriminating the pristine versus altered study sites. Raw species relative density data were used within the function *multipatt* function from the "indicspecies" package [39] that incorporated a correction for unequal group sizes. IndVal combines the information on the concentration of species abundances in a particular group (A) and the degree of occurrence in that particular group (B). Thus, ideal indicator species are those that are always present at sites in a given group and never occur in other groups [40]. The indicator values range from 0–1, the value presented in this study corresponding to the square root of the product between A and B. The significance of the indicator values was tested by 999 Monte Carlo permutations where the observed indicator value was tested against those derived from randomized data, alpha was set to 0.001.

Temporal trends of different macroinvertebrate metrics (three taxonomic, two functional and three selected taxa) were explored separately for each season as their differing taxonomic compositions (driven by organism's life-cycles) can respond differently to long-term changes in environmental conditions (e.g., [41]). Invertebrate metrics were modeled as a smoothed response over time through GAMMs via the *gamm* function in the "mgcv" package [42]. For this, cubic splines were implemented and a Gaussian distribution was modeled. Each study site was input as a random effect to account for communities from each sampling location potentially being correlated over time [8] and a first-order autoregressive process was adopted to account for temporal autocorrelation (correlated errors within years were accounted for); these criteria yielded the optimal model compared to respective models (GAMs or GAMMs) without a random effect or a first-order autoregressive process in all instances (indicated by an Akaike Information Criterion value of $\leq -2$). Then, residual diagnostics were inspected for each optimal GAMM to ensure the assumptions of homogenous variances and normal distributions were met; where these criteria were satisfied, the significance of the time smoothing parameter was obtained. In addition, periods where invertebrate responses significantly increased or decreased were explored by the first derivatives of each GAMM using the method of finite differences derived from 200 equally spaced time points over the study period [43]. When residual checks revealed the assumptions of GAMMs were not satisfied, temporal changes were instead visualised through a "LOcally Estimated Scatterplot Smoothing" (LOESS) function.

We performed all statistical analyses using R project software [25].

## 3. Results

### 3.1. Similarity among Macroinvertebrate Assemblages

The NMDS plot showed that macroinvertebrate assemblages inhabiting the River Po catchment displayed a strong geographic pattern (Figure 2). The main influence of longitude and altitude were observed along the NMDS axis 1 (Figure 2a,b) while the effect of latitude was more important along NMDS axis 2 (Figure 2c). Sites positively loaded on axis 1 tended to be characteristic of lowland sites (<400 m a.s.l.), while sites negatively loaded were typical of high-altitude sites (>1800 m a.s.l). We identified a clear altitudinal pattern in the position of the macroinvertebrate families in the ordination space (from right to left and from bottom to top Figure 2c). Assemblages inhabiting high-altitude sites were characterized mostly by families (with average NMDS1 score < −0.5) belonging to the Plecoptera order (i.e., Taeniopterygidae, Chloroperlidae, Perlodidae, Perlidae and Nemouridae), but also some Trichoptera (i.e., Glossosomatidae, Philopotamidae and Limnephilidae), Coleoptera (i.e., Scirtidae and Hydraenidae), Diptera (i.e., Blephariceridae, Thaumaleidae, Pediciidae and Psychodidae), Oligochaeta (i.e., Haplotaxidae) and Platyhelminthes (i.e., Planariidae). Macroinvertebrates inhabiting lowland sites were characterized mostly by non-insect families (with average NMDS1 score > 1.2) (e.g., Unionidae, Acroloxidae, Salifidae, Haemopidae, Corbiculidae and Hydridae), although some exceptions to this were observed with certain insect families (e.g., Sciomyzidae, Coenagrionidae and Hydrometridae). The effects of longitude and latitude on assemblage structure are less representative but coherent with the geographic characteristics of the studied catchment and partial covariates with the observed pattern of altitude.

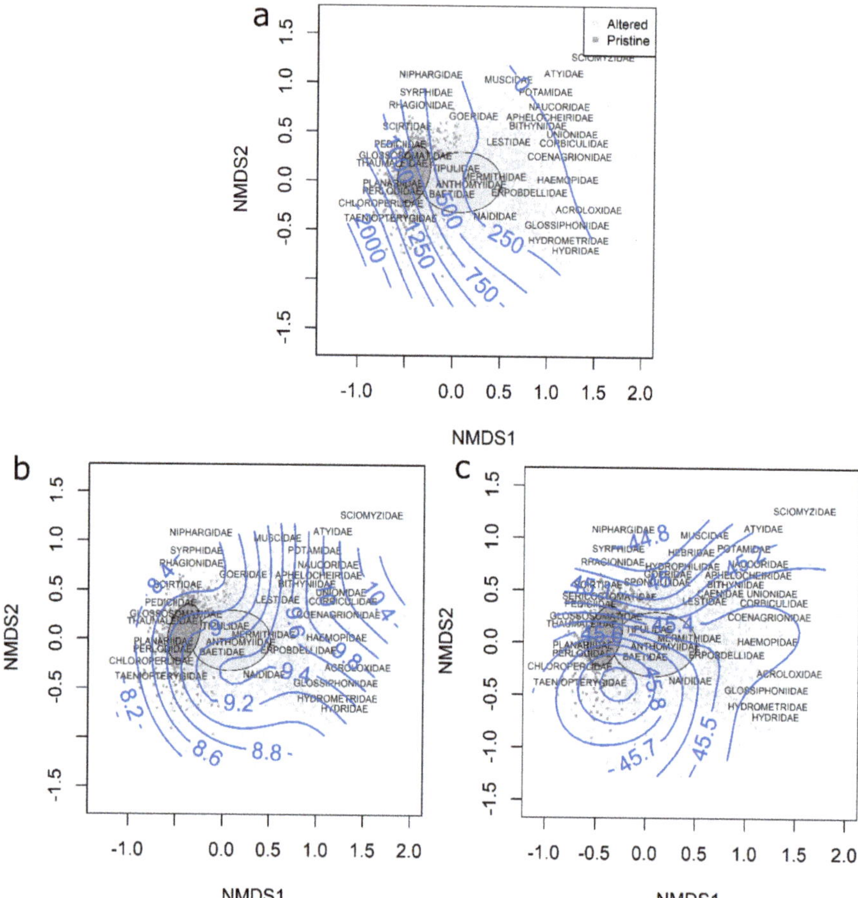

**Figure 2.** Nonmetric multidimensional scaling (NMDS) ordination plot for aquatic macroinvertebrate assemblages where the a-priori identified alteration status are colored. Dark gray denotes the assemblages that belong to pristine sites (*n* = 534) while light gray denotes those assemblages that belong to sites where an impact is present (*n* = 2563). Shaded ellipses represent the 95% confidence interval surrounding the centroid of each impact class in the ordination space. Overlaid smooth surface display blue splines using generalized additive models (GAMs) for Altitude (**a**), Longitude (**b**) and Latitude (**c**). Numbers on the splines represent respectively meter above sea level and WGS84 coordinates. Macroinvertebrate families are positioned in the ordination space with uppercase labels, as weighted averages. 3D stress = 0.16.

In the variance partitioning framework, when examining climatic, habitat and geographical variables simultaneously, climatic variables alone explained a much higher proportion of the variability than habitat or geographic variables. The total variance explained was 32.2%, with the pure effect of the climate accounting for 9.4% and, combined with the shared influence of habitat and geographic variables, this increased to 28.0% (Figure 3a). The pure effect of the habitat variables was 3.1% and, combined with the shared influence of climatic and geographic variables, this increased to 15.1%. Similarly, the pure effect of geographic variables was 1.0% and, combined with the shared influence of habitat and climatic variables, this increased to 15.1%. When examining variable types (climate, habitat and geographic) separately, the variance explained by climatic variables was mostly due to the shared influence among temperature and precipitation related variables (16.4% Figure 3b). The total variance explained by habitat was 15.1%, with the mineral substrate yielding the greatest statistical influence (pure effect 11.2%, Figure 3c). Similarly, geographic variables explained 15.1% of the statistical variation whereby altitude proved to be the most influential variable (pure effect = 8.1%, pure plus shared effect = 14.0%, Figure 3d).

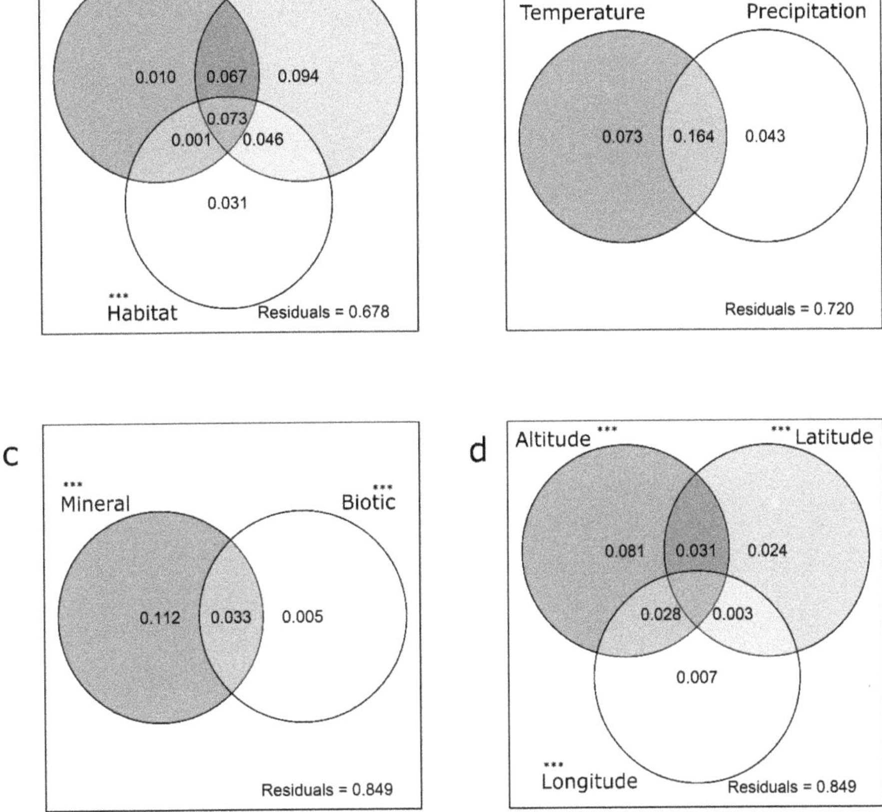


**Figure 3.** Results of variance partitioning on macroinvertebrate assemblage matrix. Panel (**a**) shows the results for the three considered groups of variables together while other panels show the detailed results for each group. Panel (**b**) refers to climatic variables, panel (**c**) to habitat variables and panel (**d**) to geographic ones. Values displayed are the adjusted $R^2$ and negative values are not shown. The unexplained portion is shown in the bottom right of panels (Residuals). Significance of the pure effects are show as asterisks. *** $p < 0.001$.

### 3.2. Presence of Human Impacts

PERMANOVA highlighted significant differences among the assemblages inhabiting pristine sites versus those subjected to some form of human alteration ($F = 167.97$, $R^2 = 0.05$, $p < 0.001$). The two clusters cover most of the altitudinal and latitudinal ranges of the River Po catchment, while the pristine sites were better represented in its western part (Figure S1).

Indicator species analysis highlighted a range of macroinvertebrate families that differed significantly between the two alteration classes, with many insect families associated with pristine sites and non-insect families associated with altered sites (Table 1). Twenty-three macroinvertebrate families, spanning across several taxonomic orders, displayed greater affinities for pristine sites, while 12 families showed greater affinity for the sites impacted by anthropogenic environmental alterations (Table 1). For all the selected families, the A term was bigger than the B term, suggesting these taxa are good indicators of the altered group, but are present only occasionally.

**Table 1.** Results of the IndVal analysis for Pristine and Altered sites. A represent the probability that the surveyed site belongs to the target site group given the fact that the family has been found, whereas B represent the probability of finding the family in sites belonging to the site group. All the reported associations were significant after 999 permutation at alpha = 0.001.

| Taxa Group | Family | A | B | IndVal |
|---|---|---|---|---|
| **Pristine Sites** | | | | |
| Oligochaeta | Haplotaxidae | 0.76 | 0.90 | 0.82 |
| Plecoptera | Chloroperlidae | 0.80 | 0.70 | 0.75 |
| Trichoptera | Polycentropodidae | 0.76 | 0.71 | 0.74 |
| Diptera | Dixidae | 0.71 | 0.73 | 0.72 |
| Ephemeroptera | Leptophlebiidae | 0.80 | 0.59 | 0.69 |
| Plecoptera | Taeniopterygidae | 0.76 | 0.52 | 0.63 |
| Oligochaeta | Enchytraeidae | 0.78 | 0.49 | 0.62 |
| Diptera | Psychodidae | 0.76 | 0.46 | 0.59 |
| Diptera | Rhagionidae | 0.70 | 0.48 | 0.58 |
| Plecoptera | Perlodidae | 0.86 | 0.32 | 0.52 |
| Trichoptera | Philopotamidae | 0.82 | 0.31 | 0.51 |
| Trichoptera | Glossosomatidae | 0.75 | 0.32 | 0.49 |
| Trichoptera | Beraeidae | 0.71 | 0.28 | 0.44 |
| Coleoptera | Scirtidae | 0.80 | 0.20 | 0.40 |
| Trichoptera | Odontoceridae | 0.79 | 0.20 | 0.40 |
| Diptera | Thaumaleidae | 0.69 | 0.19 | 0.36 |
| Plecoptera | Perlidae | 0.83 | 0.16 | 0.36 |
| Diptera | Blephariceridae | 0.81 | 0.16 | 0.36 |
| Platyhelminthes | Planariidae | 0.83 | 0.15 | 0.35 |
| Diptera | Pediciidae | 0.72 | 0.16 | 0.34 |
| Diptera | Athericidae | 0.84 | 0.09 | 0.28 |
| Coleoptera | Hydraenidae | 0.90 | 0.04 | 0.19 |
| Plecoptera | Nemouridae | 0.96 | 0.02 | 0.15 |
| **Altered sites** | | | | |
| Oligochaeta | Tubificidae | 0.80 | 0.70 | 0.59 |
| Hirudinea | Erpobdellidae | 0.76 | 0.71 | 0.55 |
| Crustacea | Gammaridae | 0.71 | 0.73 | 0.51 |
| Crustacea | Asellidae | 0.80 | 0.59 | 0.41 |
| Gastropoda | Physidae | 0.76 | 0.52 | 0.37 |
| Odonata | Calopterygidae | 0.78 | 0.49 | 0.30 |
| Gastropoda | Lymnaeidae | 0.76 | 0.46 | 0.30 |
| Hirudinea | Glossiphoniidae | 0.70 | 0.48 | 0.27 |
| Bivalvia | Corbiculidae | 0.86 | 0.32 | 0.27 |
| Heteroptera | Aphelocheiridae | 0.82 | 0.31 | 0.24 |
| Gastropoda | Hydrobiidae | 0.75 | 0.32 | 0.23 |
| Gastropoda | Planorbidae | 0.71 | 0.28 | 0.20 |

*3.3. Metrics and Indices*

The density of macroinvertebrates ranged between 12 and 36,820 individuals *m$^{-2}$ (1673 ± 2260, mean ± standard deviation) in the whole study area, whereas family richness varies between 3 and 44 (19 ± 7). The macroinvertebrate families selected on the basis of their wide geographic distribution and relevance to bioassessment (i.e., Nemouridae, Heptageniidae and Limnephilidae) were often present in the considered samples (48%, 77% and 42%, respectively) and their density ranged between 0–830,

0–1660 and 0–6760 (respectively) per square meter. The percentage of Ephemeroptera, Plecoptera and Trichoptera individuals (%EPT) within a sample was 51 ± 25%, while the Shannon diversity index (H′) ranged between 0.01 to 2.96. The Pearson's correlation between metrics and indices was significant in the majority of instances (Figure 4), but the only two pairs of highly correlated indices (r > 0.70) were taxonomic richness versus functional richness (r = 0.76, $p < 0.001$) and H′ versus functional diversity (r = 0.80, $p < 0.001$).

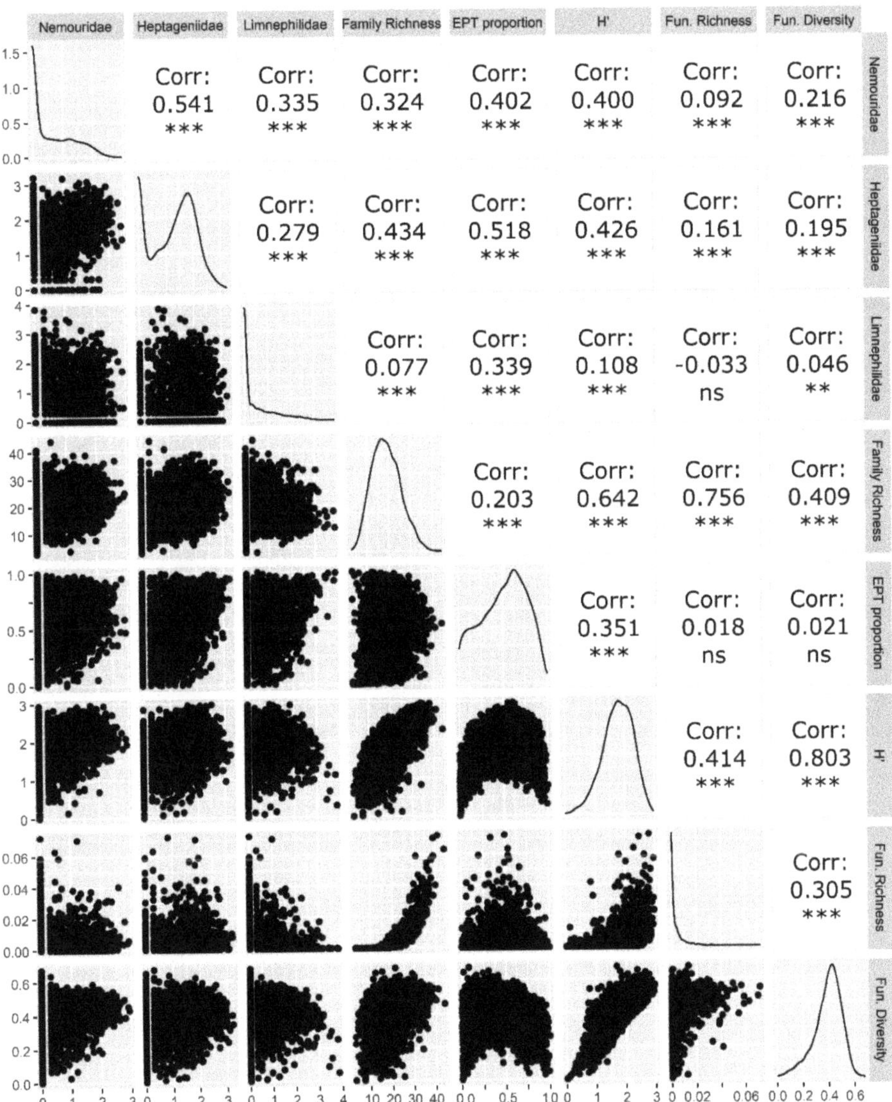

**Figure 4.** Correlation among macroinvertebrate communities metrics and indices. The diagonal represents the density function of each metric or index, below the diagonal the pairs plot are presented and above the diagonal the Pearson's correlation between each pair of metrics and indices is reported along with the significance levels. Levels of statistical significance are as follows: ***, $p < 0.001$; **, $p < 0.01$; ns, $p > 0.05$.

In the variance partitioning framework, the relative density of the three selected macroinvertebrate families (Nemouridae, Heptageniidae and Limnephilidae) and functional diversity were less influenced by the geographic position, climate and habitat than the multivariate assemblage structure and other univariate metrics (Table 2). The total variance explained for structural indices and functional richness ranged between 27 and 31% and the influence of climatic variables was always the greatest (24–28%), as observed also for the multivariate assemblage structure. The variance explained by climatic variables was mostly due to the shared influence with habitat and geographic position than to its pure effect for all metrics (Figure S2).

**Table 2.** Results of variance partitioning on macroinvertebrate metrics and indices. Values reported are the adjusted $R^2$.

| Metric/Index | Total Explained Variance | Geographic | Climatic | Habitat |
|---|---|---|---|---|
| Assemblage distance matrix | 32.2 | 15.1 | 28.0 | 15.5 |
| Nemouridae density | 16.2 | 13.2 | 14.1 | 5.4 |
| Heptageniidae density | 10.3 | 4.9 | 8.4 | 3.6 |
| Limnephilidae density | 6.0 | 2.8 | 5.9 | 0.1 |
| Family richness | 31.3 | 8.6 | 27.6 | 11.0 |
| EPT proportion | 30.2 | 13.8 | 25.7 | 19.8 |
| Shannon | 27.6 | 7.6 | 24.6 | 11.3 |
| Functional Richness | 26.7 | 13.9 | 24.6 | 6.6 |
| Functional Diversity | 11.3 | 4.4 | 9.4 | 2.5 |

*3.4. Temporal Trends of Metrics and Indices*

Using GAMMs, it was possible to assess the significance of the temporal trends for four out of eight tested metrics and all proved to be significant ($p < 0.05$) for all the seasons with coherent patterns. Over the first few years of monitoring (*c.* 2008–2011), family richness, Shannon and functional diversity declined significantly (Figure 5a–c,e). The taxonomic richness and Shannon notably increased between *c.* 2011–2017, decreased significantly thereafter (Figure 5a–c,g–i); these patterns were broadly mirrored by the functional diversity response, although fewer significant changes were detected for this metric (Figure 5j–l). The %EPT broadly contradicted the temporal trends observed for the other metrics, although they largely displayed a lack of significant temporal variations. One exception to this was a significant increase in values in the fall months after 2017, whereby an abrupt decline (*c.* 10–15%) occurred across the two years prior and then completely recovered by the end of 2018.

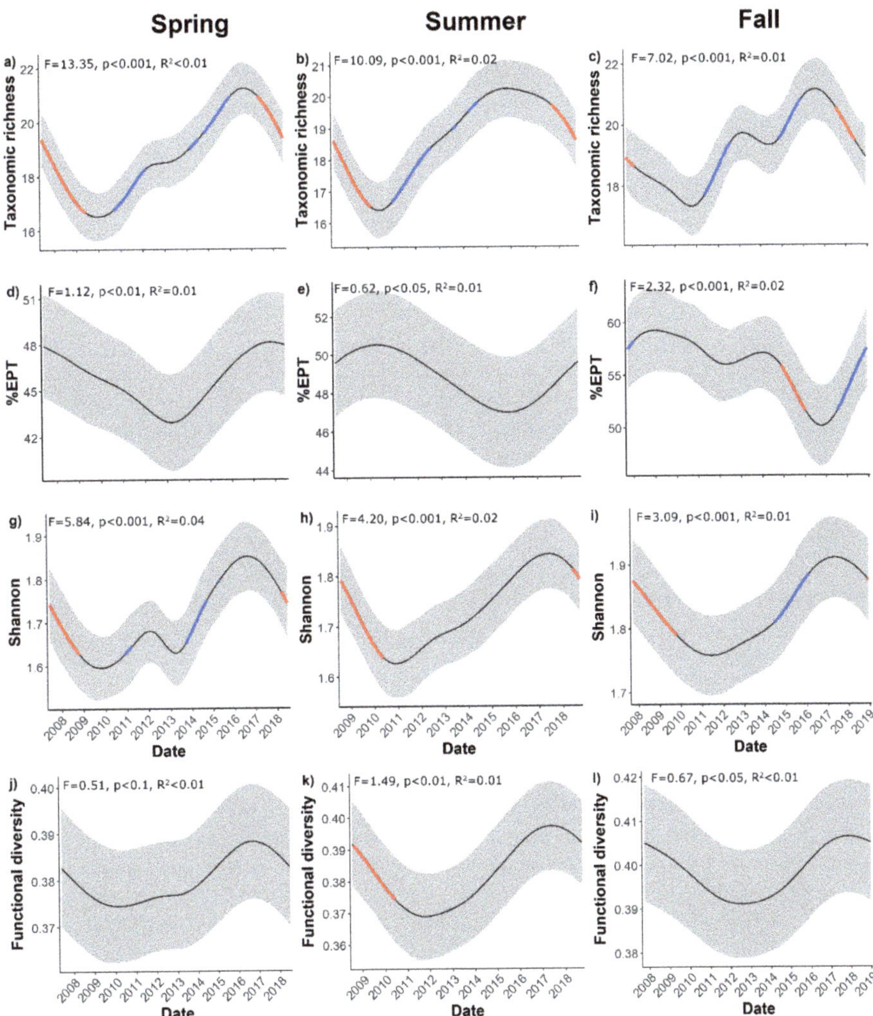

**Figure 5.** Temporal trajectories of (**a–c**) family richness, (**d–f**) %EPT, (**g–i**) Shannon diversity index (H′) and (**j–l**) functional diversity as a function of the generalized additive mixed-effect models (GAMMs) outputs for the different seasons. The significance of the time smoothing parameters is reported within each panel. Significant increases or decreases in slope are represented respectively in blue and red colors. Corresponding graphs, only to visualize temporal trends via LOESS (Locally Estimated Scatterplot Smoothing) for the other metrics and indices are shown in Figure S3.

## 4. Discussion

Findings from this study highlighted that the structure of macroinvertebrate assemblages in the River Po catchment (Northern Italy) were controlled by multiple drivers acting at both coarse (e.g., altitude and climate) [44] and local spatial scales (e.g., habitat characteristics) [45]. The variance partitioning analyses highlighted that the variation in macroinvertebrate assemblages was mainly explained by the shared influence of the three groups of environmental variables and indicated that climate defined most of the variance explained, followed by habitat and then geographic position. Altitude has a well-documented control of flow and thermal regimes, which are widely recognized as primary determinants shaping macroinvertebrate assemblage structures (e.g., [46–48]). Mineral substrate composition showed greater association with macroinvertebrate than biological substrates, which is likely due to the latter being almost completely absent in sites above 500 m a.s.l. [21], while coarse substrates are commonly associated with steeper channel slope, and hence occur more widely at higher altitudes. Our findings highlighted that altitude was the main geographic driver within the study area, as reported in many other studies (e.g., [49,50]), thus confirming our H1 hypothesis.

The results of NMDS analyses showed that the macroinvertebrate assemblages of the present study were strongly driven by geographic variables and particularly by altitude, confirming again our $H_1$ hypothesis. High altitude assemblages, as well as those inhabiting pristine sites, were dominated by Plecoptera families. The association between Plecoptera and altitude is well known (e.g., [51,52]), as is their vulnerability to climatic changes through their restricted altitudinal niche, which has been previously highlighted [53] and represents a new threat for this order in the study region (alongside pollution) [54]. Families belonging to other orders such as Trichoptera, Coleoptera, Diptera, Oligochaeta and Platyhelminthes showed similar altitudinal gradients (i.e., strong association with high-altitude sites). These results are due, at least partially, to their biogeographic distribution and not the effect of human alterations. Among the families representative of lowland sites, some are commonly associated with the potamal stretches (longitudinal distribution according to [30]) of rivers and streams (e.g., Unionidae, Corbiculidae), or with lentic waters (e.g., Hydrometridae, Acroloxidae), whereas others are more habitat generalist (e.g., Salifidae, Haemopidae). While some of these are considered indicators of good environmental status (e.g., Unionidae and Hydrometridae), Corbiculidae is considered an invasive family in Italy and was associated with altered sites in this work. In summary, high altitude assemblages were dominated by insects, while lowland assemblages were dominated by non-insects, and were more susceptible to biological invasions [55].

According to PERMANOVA, significant differences occurred among the assemblages inhabiting the pristine versus altered sites, confirming our $H_2$ hypothesis. Moreover, IndVal analysis identified various families associated with the different alteration status, supporting our $H_3$ hypothesis. The macroinvertebrate families associated with pristine sites were typically indicators of good water quality, as highlighted within Biological Monitoring Working Party (BMWP—[13]). Conversely, most families associated with altered sites are generally classified indicators of impaired water quality. Such findings agree with the extensive literature available about the tolerance of macroinvertebrate families to water quality impairment (e.g., [56–58]); nonetheless, many families commonly classified as sensitive to nutrient enrichment (e.g., Leuctridae, Heptageniidae and Leptoceridae) did not display significant differences between their abundance in pristine and altered sites. In addition, many of the families associated with pristine sites are classified as rheophilic within the Lotic-invertebrate Index for Flow Evaluation metric (LIFE—[14]), which is widely used to define hydroecological relationships. Conversely, many of the families associated with altered sites possess LIFE scores that indicate they are reflective of slower flow conditions. Such ecological trends are critical within biomonitoring assessments, and as such further explorations of the relationships among selected pressures and the responses of biological indices required to help guide specific management actions [59].

The results of the variance partitioning analyses applied to structural indices, functional richness and those of multivariate assemblage structure were broadly comparable; while the absolute abundance of Nemouridae, Heptageniidae and Limnephilidae and the functional diversity were less responsive to the considered environmental drivers. We processed a limited number of metrics which responded differently to different stressors. Given that individually, these metrics only characterize a small part of the assemblage, they may lack sensitivity to changes in certain environmental conditions or when multiple anthropogenic stressors are present [60], which may have resulted in incongruent responses and certain ecological trends not being detected [61]. Future studies that focus on the effect of specific anthropogenic modifications in this catchment (as well as studies using biomonitoring data elsewhere) could build upon this study and focus on macroinvertebrate indices and metrics developed to assess a species stressor (e.g., BMWP for nutrient enrichment and LIFE for hydrological alterations).

Notwithstanding, univariate indicators used in this study consider complete macroinvertebrate assemblages which reliably characterize assemblage structure and are capable of capturing ecological responses to environmental gradients. The relative density of selected families are likely controlled both by stochasticity and by natural and anthropogenic drivers that act at smaller temporal and spatial scales than considered in this research, such as short-term climate variations and/or local pollution events [62].

In accordance with our last hypothesis ($H_4$) we identified significant temporal trends for some of the studied metrics. Inter-annual variations in macroinvertebrate assemblages can be due to environmental changes such as climatic variations driving river flow and thermal regimes [63], changes in anthropogenic pressures (e.g., river restoration), to the establishment of invasive species [8], or to the interactive effects of those drivers [41]. The first few years of the study period witnessed evident declines in various response metrics; namely, taxonomic richness and diversity, during a time when discharges were increasing following a major drought in 2007 [64]. Comparably, significant increases in the same metrics were observed in 2017, when an unprecedent heatwave occurred across southern Europe that triggered a severe drought in many areas, persisting into fall [65]. These significant temporal trends indicate that macroinvertebrate assemblages inhabiting the Po catchment thrive during low-flow periods, likely due to many taxa within the regions displaying strong tolerance mechanisms to drought in a region known for its harsh, dry summers. Despite this, Ephemeroptera, Plecoptera and Trichoptera taxa displayed the opposite trend, most notably during autumn months. Such findings are unsurprising, as at the order level these taxa are negatively associated with low-flow periods [66], while autumn months depicts a critical time for the larval recruitment development. As such, significant temporal variations displayed in the Po catchment are most likely attributed to inter-annual flow and thermal regime variations. It is unlikely that temporal trends within this large catchment were driven by modifications of the anthropogenic pressures because those would have varied spatially across the Po catchment and generally do not present abrupt changes like the ones that we observed for the macroinvertebrate assemblages [67].

## 5. Conclusions

### *Final Remarks*

Our results provide evidence of the link between local climate and macroinvertebrate assemblages in the catchment of River Po (North Italy). The detail of this study dataset allowed for important environmental attributes (e.g., altitude, habitat characteristics) shaping biotic communities to be identified, along with ecologically important and vulnerable regions and time periods (e.g., extreme climatic events). Such research is required globally to help inform large-scale management and restoration efforts that are sustainable over long-term periods. While this study represents the first of its kind across the region, further studies need to focus both on the effects of inter-annual climate variability on macroinvertebrate assemblages to further explore temporal trends described here, as well as to cultivate a better understanding of primary stressors shaping biotic assemblages through the use of specifically developed metrics (e.g., nutrient enrichment—BMWP; flow—LIFE,). To achieve a better understanding of macroinvertebrate assemblage dynamics within this catchment, the biomonitoring network can be improved by focusing on "reference" sites (largely free from human interference), although this study highlighted the difficulty finding such conditions in lowland environments. Further analyses should focus on the response of the families which we have identified representative of pristine sites in this study; in medium-high altitude sites in particular, such research could unveil differences in ecological dynamics between pristine and impacted sites. Moreover, from a management perspective, there remains a global need to incorporate measures of different anthropogenic impacts within (or alongside) ecological databases, as this would allow key drivers and stressors to be identified and integrated into effective management strategies.

**Supplementary Materials:** The following are available online at http://www.mdpi.com/2073-4441/12/9/2452/s1, Table S1: Macroinvertebrate functional traits examined within this study, with biological traits in non-italicized text and ecological traits being italicized, Figure S1: Distribution of pristine and altered sites along the geographic gradients, Figure S2: Results of variance partitioning on macroinvertebrate metrics and indices. Values displayed are the adjusted $R^2$ and negative values are not shown. The unexplained portion is shown in the bottom right of panels (Residuals), Figure S3: Temporal trajectories of (a–c) $\log_{10}$ transformed Nemouridae absolute abundance, (d–f) $\log_{10}$ transformed Heptageniidae absolute abundance, (g–i) $\log_{10}$ transformed Limnephilidae absolute abundance and (j–l) Functional Richness visualized using LOESS (Locally Estimated Scatterplot Smoothing) for the different seasons.

**Author Contributions:** Conceptualization, R.F., A.B. and A.L.; Methodology, R.F. and J.C.W.; Software, R.F., J.C.W. and A.L.; Formal Analysis, R.F. and J.C.W.; Data Curation, R.F.; Writing—Original Draft Preparation, R.F., A.B., J.C.W. and A.L.; Visualization, R.F. and J.C.W.; Supervision, A.B. and A.L.; Writing—review and editing, R.F., A.B., J.C.W. and A.L. All authors have read and agreed to the published version of the manuscript.

**Funding:** This research received no external funding.

**Acknowledgments:** The Authors would like to thank all technicians and operators of the environmental agencies for the sharing of data collected during water quality monitoring programs over the past decade, without which this work would not have been possible. We also thank the two anonymous reviewers for their constructive comments on an earlier version of this paper.

**Conflicts of Interest:** The authors declare no conflict of interest.

## References

1. Vugteveen, P.; Leuven, R.S.E.W.; Huijbregts, M.A.J.; Lenders, H.J.R. Redefinition and Elaboration of River Ecosystem Health: Perspective for River Management. *Hydrobiologia* **2006**, *565*, 289–308. [CrossRef]

2. Schmutz, S. *Riverine Ecosystem Management*; Springer International Publishing: New York, NY, USA, 2018; ISBN 9783319732497.

3. Wohl, E.; Lane, S.; Wilcox, A. The Science and Practice of River Restoration. *Water Resour. Res.* **2015**, *51*, 5974–5997. [CrossRef]

4. Vörösmarty, C.J.; McIntyre, P.B.; Gessner, M.O.; Dudgeon, D.; Prusevich, A.; Green, P.; Glidden, S.; Bunn, S.E.; Sullivan, C.A.; Liermann, C.R.; et al. Global threats to human water security and river biodiversity. *Nature* **2010**, *467*, 555–561. [CrossRef] [PubMed]

5. Allan, J.D.; Flecker, A.S. Biodiversity Conservation in Running Waters. *Bioscience* **1993**, *43*, 32–43. [CrossRef]

6. Lewin, I.; Jusik, S.; Szoszkiewicz, K.; Czerniawska-Kusza, I.; Ławniczak, A.E. Application of the new multimetric MMI_PL index for biological water quality assessment in reference and human-impacted streams (Poland, the Slovak Republic). *Limnologica* **2014**, *49*, 42–51. [CrossRef]

7. Nilsson, C.; Berggren, K. Alterations of riparian ecosystemes caused by river regulation. *Bioscience* **2000**, *50*, 783–792. [CrossRef]

8. Mathers, K.L.; White, J.C.; Fornaroli, R.; Chadd, R. Flow regimes control the establishment of invasive crayfish and alter their effects on lotic macroinvertebrate communities. *J. Appl. Ecol.* **2020**, *57*, 886–902. [CrossRef]

9. EC European Community. *Directive 2000/60/EC of the European Parliament and of the Council of 23 October 2000 Establishing a Framework for Community Action in the Field of Water Policy*; European Community: Brussels, Belgium, 2000; Volume L327/1.

10. Ofenböck, T.; Moog, O.; Gerritsen, J.; Barbour, M. A stressor specific multimetric approach for monitoring running waters in Austria using benthic macro-invertebrates. *Hydrobiologia* **2004**, *516*, 251–268. [CrossRef]

11. Hering, D.; Moog, O.; Sandin, L.; Verdonschot, P.F.M. Overview and application of the AQEM assessment system. *Hydrobiologia* **2004**, *516*, 1–20. [CrossRef]

12. Pinto, P.; Rosado, J.; Morais, M.; Antunes, I. Assessment methodology for southern siliceous basins in Portugal. *Hydrobiologia* **2004**, *516*, 191–214. [CrossRef]

13. Armitage, P.D.; Moss, D.; Wright, J.F.; Furse, M.T. The performance of a new biological water quality score system based on macroinvertebrates over a wide range of unpolluted running-water. *Water Res.* **1983**, *17*, 333–347. [CrossRef]

14. Extence, C.; Balbi, D.M.; Chadd, R.P. River flow indexing using British benthic macroinvertebrates: A framework for setting hydroecological objectives. *Regul. Rivers Res. Manag.* **1999**, *15*, 543–574. [CrossRef]

15. Extence, C.; Chadd, R.P.; England, J.; Dunbar, M.J.; Wood, P.J.; Taylor, E.D. The assessment of fine sediment accuulation in rivers using macroinvertebrate community response. *River Res. Appl.* **2013**, *29*, 17–55. [CrossRef]

16. Suren, A.M.; Jowett, I.G. Effects of floods versus low flows on invertebrates in a New Zealand gravel-bed river. *Freshw. Biol.* **2006**, *51*, 2207–2227. [CrossRef]

17. Merritt, R.W.; Wallace, J.; Higgins, M.; Alexander, M.; Berg, M.; Morgan, W.; Cummins, K.; Vandeneeden, B. Procedures for the functional analysis of invertebrate communities of the kissimmee River-floodplain ecosystem. *Fla. Sci.* **1996**, *59*, 216–274.

18. Reynoldson, T.B.; Norris, R.H.; Resh, V.H.; Day, K.E.; Rosenberg, D.M. The reference condition: A comparison of multimetric and multivariate approaches to assess water-quality impairment using benthic macroinvertebrates. *J. N. Am. Benthol. Soc.* **1997**, *16*, 833–852. [CrossRef]

19. Erba, S.; Cazzola, M.; Belfiore, C.; Buffagni, A. Macroinvertebrate metrics responses to morphological alteration in Italian rivers. *Hydrobiologia* **2020**, *847*, 2169–2191. [CrossRef]

20. Calabrese, S.; Mezzanotte, V.; Marazzi, F.; Canobbio, S.; Fornaroli, R. The influence of multiple stressors on macroinvertebrate communities and ecosystem attributes in Northern Italy pre-Alpine rivers and streams. *Ecol. Indic.* **2020**, *115*, 106408. [CrossRef]

21. Fornaroli, R.; Agostini, A.; Arnaud, E.; Berselli, A.; Bettoni, E.; Boggero, A.; Brolandelli, C.; Cadrobbi, G.; Cason, M.; Castelli, L.; et al. A ten-year geographic data set on the occurrence and abundance of macroinvertebrates in the River Po basin (Italy). *Biogeographia* **2020**, (in press).

22. *ISPRA Istituto Superiore per la Protezione e Ricerca Ambientale. Manuali e Linee Guida 107/2014. Report in Italian*; ISPRA Linee guida per la valutazione della componente macrobentonica fluviale ai sensi del DM 260/2010; ISPRA: Roma, Italy, 2014; p. 88.

23. EEA European Environment Agency. Biogeographical Regions in Europe. Internet Database. Available online: https://www.eea.europa.eu/data-and-maps/figures/biogeographical-regions-in-europe-2 (accessed on 31 August 2020).

24. Laini, A.; Bolpagni, R.; Daniel, B.; Burgazzi, G.; Guareschi, S.; Cédric, M.; Tommaso, C. biomonitoR: An R package for calculating biomonitoring indices of running waters. XXVIII Congr. Ital. Ecol. Soc. 2018. Available online: https://www.biomonitor.it/ (accessed on 31 August 2020).

25. R Core Team. *R: A Language and Environment for Statistical Computing*; R Foundation for Statistical Computing: Vienna, Austria, 2019.

26. Hijmans, R.J. Raster: Geographic Data Analysis and Modeling. R package Version 3.0-12. Available online: https://CRAN.R-project.org/package=raster (accessed on 31 August 2020).

27. Schmidt-Kloiber, A.; Bremerich, V.; De Wever, A.; Jähnig, S.C.; Martens, K.; Strackbein, J.; Tockner, K.; Hering, D. The Freshwater Information Platform: A global online network providing data, tools and resources for science and policy support. *Hydrobiologia* **2019**, *838*, 1–11. [CrossRef]

28. Schmidt-Kloiber, A.; Hering, D. www.freshwaterecology.info—An online tool that unifies, standardises and codifies more than 20,000 European freshwater organisms and their ecological preferences. *Ecol. Indic.* **2015**, *53*, 271–282. [CrossRef]

29. Schmera, D.; Podani, J.; Heino, J.; Erös, T.; Poff, N.L.R. A proposed unified terminology of species traits in stream ecology. *Freshw. Sci.* **2015**, *34*, 823–830. [CrossRef]

30. Tachet, H.; Richoux, P.; Bournaud, M.; Usseglio-Polatera, P. *Invertébrés d'eau Douce*, 2nd ed.; CNRS Editions: Paris, France, 2010; ISBN 2-271-05745-0.

31. Villéger, S.; Mason, N.W.H.; Mouillot, D. New multidimensional functional diversity indices for a multifaceted framework in functional ecology. *Ecology* **2008**, *89*, 2290–2301. [CrossRef] [PubMed]

32. Rao, R.C. Diversity and Dissimilarity Coefficients: A Unified Approach. *Theor. Popul. Biol.* **1982**, *21*, 24–43. [CrossRef]

33. Botta-Dukát, Z. Rao's quadratic entropy as a measure of functional diversity based on multiple traits. *J. Veg. Sci.* **2005**, *16*, 533–540. [CrossRef]

34. Clarke, K.R.; Ainsworth, M. A method of linking multivariate community structure to environmental variables. *Mar. Ecol. Prog. Ser.* **1993**, *92*, 205–219. [CrossRef]

35. Oksanen, J.; Blanchet, F.G.; Friendly, M.; Kindt, R.; Legendre, P.; Mcglinn, D.; Minchin, P.R.; O'Hara, R.B.; Simpson, G.L.; Solymos, P.; et al. *Vegan: Community Ecology Package*; R package Version 2.5-6. Available online: https://CRAN.R-project.org/package=vegan (accessed on 31 August 2020).

36. Tonkin, J.D. Drivers of macroinvertebrate community structure in unmodified streams. *PeerJ* **2014**, *2*, e465. [CrossRef] [PubMed]

37. Borcard, D.; Legendre, P.; Drapeau, P. Partialling out the Spatial Component of Ecological Variation. *Ecology* **1992**, *73*, 1045–1055. [CrossRef]

38. Peres-Neto, P.R.; Legendre, P.; Dray, S.; Borcard, D. Variance partitioning of species data matrices: Estimation and comparison of fractions. *Ecology* **2006**, *87*, 2614–2625. [CrossRef]

39. De Cáceres, M.; Legendre, P. Associations between species and groups of sites: Indices and statistical inference. *Ecology* **2009**, *90*, 3566–3574. [CrossRef]

40. Dufrêne, M.; Legendre, P. Species assemblages and indicator species: The need for a flexible asymmetrical approach. *Ecol. Monogr.* **1997**, *67*, 345–366. [CrossRef]

41. Durance, I.; Ormerod, S.J. Trends in water quality and discharge confound long-term warming effects on river macroinvertebrates. *Freshw. Biol.* **2009**, *54*, 388–405. [CrossRef]

42. Wood, S.N. *Package 'mgcv'*. R package Version 1.8-33. Available online: https://CRAN.R-project.org/package= vegan (accessed on 31 August 2020).

43. Monteith, D.T.; Evans, C.D.; Henrys, P.A.; Simpson, G.L.; Malcolm, I.A. Trends in the hydrochemistry of acid-sensitive surface waters in the UK 1988–2008. *Ecol. Indic.* **2014**, *37*, 287–303. [CrossRef]

44. Vaughan, I.P.; Ormerod, S.J. Large-scale, long-term trends in British river macroinvertebrates. *Glob. Chang. Biol.* **2012**, *18*, 2184–2194. [CrossRef]

45. Lamouroux, N.; Dolédec, S.; Gayraud, S. Biological traits of stream macroinvertebrate communities: Effects of microhabitat, reach, and basin filters. *J. N. Am. Benthol. Soc.* **2004**, *23*, 449–466. [CrossRef]

46. Jackson, H.M.; Gibbins, C.N.; Soulsby, C. Role of discharge and temperature variation in determining invertebrate community structure in a regulated river. *River Res. Appl.* **2007**, *23*, 651–669. [CrossRef]

47. Chinnayakanahalli, K.J.; Hawkins, C.P.; Tarboton, D.G.; Hill, R.A. Natural flow regime, temperature and the composition and richness of invertebrate assemblages in streams of the western United States. *Freshw. Biol.* **2011**, *56*, 1248–1265. [CrossRef]

48. White, J.C.; Hannah, D.M.; House, A.; Beatson, S.J.V.; Martin, A.; Wood, P.J. Macroinvertebrate responses to flow and stream temperature variability across regulated and non-regulated rivers. *Ecohydrology* **2017**, *10*, e1773. [CrossRef]

49. Lewin, I.; Szoszkiewicz, K.; Jusik, S.; Ławniczak, A.E. Influence of selected environmental factors on macroinvertebrates in mountain streams. *Open Life Sci.* **2015**, *10*, 99–111. [CrossRef]

50. Suren, A.M. Macroinvertebrate communities of streams in western Nepal: Effects of altitude and land use. *Freshw. Biol.* **1994**, *32*, 323–336. [CrossRef]

51. Kamler, E. Distribution of Plecoptera and Ephemeroptera in relation to altitude above mean sea level and current speed in mountain waters. *Pol. Arch. Hidrobiol.* **1967**, *2*, 29–42.

52. Hynes, H.B.N. Biology of Plecoptera. *Annu. Rev. Entomol.* **1976**, *21*, 135–153. [CrossRef]

53. De Figueroa, J.M.T.; López-Rodríguez, M.J.; Lorenz, A.; Graf, W.; Schmidt-Kloiber, A.; Hering, D. Vulnerable taxa of European Plecoptera (Insecta) in the context of climate change. *Biodivers. Conserv.* **2010**, *19*, 1269–1277. [CrossRef]

54. Fochetti, R. Endemism in the Italian stonefly-fauna (Plecoptera). *Zootaxa* **2020**, *4722*, 381–388. [CrossRef] [PubMed]

55. Fenoglio, S.; Bonada, N.; Guareschi, S.; López-Rodríguez, M.J.; Millán, A.; Tierno De Figueroa, J.M. Freshwater ecosystems and aquatic insects: A paradox in biological invasions. *Biol. Lett.* **2016**, *12*, 20151075. [CrossRef] [PubMed]

56. Paisley, M.F.; Trigg, D.J.; Walley, W.J. Revision of the Biological Monitoring Working Party (BMWP) score system: Derivation of present-only and abundance-related scores from field data. *River Res. Appl.* **2014**, *30*, 887–904. [CrossRef]

57. Wilding, N.A.; White, J.C.; Chadd, R.P.; House, A.; Wood, P.J. The influence of flow permanence and drying pattern on macroinvertebrate biomonitoring tools used in the assessment of riverine ecosystems. *Ecol. Indic.* **2018**, *85*, 548–555. [CrossRef]

58. Guareschi, S.; Laini, A.; Sánchez-Montoya, M.M. How do low-abundance taxa affect river biomonitoring? Exploring the response of different macroinvertebrate-based indices. *J. Limnol.* **2017**, *76*, 9–20. [CrossRef]

59. Clews, E.; Ormerod, S.J. Improving bio-diagnostic monitoring using simple combinations of standard biotic indices. *River Res. Appl.* **2009**, *25*, 348–361. [CrossRef]

60. Laini, A.; Bolpagni, R.; Cancellario, T.; Guareschi, S.; Racchetti, E.; Viaroli, P. Testing the response of macroinvertebrate communities and biomonitoring indices under multiple stressors in a lowland regulated river. *Ecol. Indic.* **2018**, *90*, 47–53. [CrossRef]

61. Yuan, L.L.; Norton, S.B. Comparing responses of macroinvertebrate metrics to increasing stress. *J. N. Am. Benthol. Soc.* **2003**, *22*, 308–322. [CrossRef]

62. Robinson, C.T.; Minshall, G.W.; Royer, T.V. Inter-annual patterns in macroinvertebrate communities of wilderness streams in Idaho, U.S.A. *Hydrobiologia* **2000**, *421*, 187–198. [CrossRef]

63. Monk, W.A.; Wood, P.J.; Hannah, D.M.; Wilson, D.A. Macroinvertebarte community response to inter-annual and regional river flow regime dynamics. *River Res. Appl.* **2008**, *245*, 988–1001. [CrossRef]

64. Montanari, A. Hydrology of the Po River: Looking for changing patterns in river discharge. *Hydrol. Earth Syst. Sci.* **2012**, *16*, 3739–3747. [CrossRef]

65. Kew, S.F.; Philip, S.Y.; Van Oldenborgh, G.J.; Otto, F.E.L.; Vautard, R.; Van Der Schrier, G. The exceptional summer heat wave in southern Europe 2017. *Bull. Am. Meteorol. Soc.* **2019**, *100*, S49–S53. [CrossRef]

66. Datry, T.; Larned, S.T.; Fritz, K.M.; Bogan, M.T.; Wood, P.J.; Meyer, E.I.; Santos, A.N. Broad-scale patterns of invertebrate richness and community composition in temporary rivers: Effects of flow intermittence. *Ecography* **2014**, *37*, 94–104. [CrossRef]

67. Randin, C.F.; Ashcroft, M.B.; Bolliger, J.; Cavender-Bares, J.; Coops, N.C.; Dullinger, S.; Dirnböck, T.; Eckert, S.; Ellis, E.; Fernández, N.; et al. Monitoring biodiversity in the Anthropocene using remote sensing in species distribution models. *Remote Sens. Environ.* **2020**, *239*, 111626. [CrossRef]

*Article*

# Aquatic Insects and Benthic Diatoms: A History of Biotic Relationships in Freshwater Ecosystems

Stefano Fenoglio [1,4,*], José Manuel Tierno de Figueroa [2], Alberto Doretto [3,4], Elisa Falasco [1,4] and Francesca Bona [1,4]

1    Department of Life Sciences and Systems Biology, Università degli Studi di Torino, Via A. Albertina, 13, 10123 Torino, Italy; elisa.falasco@unito.it (E.F.); francesca.bona@unito.it (F.B.)
2    Department of Zoology, University of Granada, Campus Fuentenueva s/n, 18071 Granada, Spain; jmtdef@ugr.es
3    Department of Sciences and Technological Innovation, Università degli Studi del Piemonte Orientale, Viale Teresa Michel, 11-15121 Alessandria, Italy; alberto.doretto@uniupo.it
4    ALPSTREAM—Alpine Stream Research Center, 12030 Ostana (CN), Italy
*    Correspondence: stefano.fenoglio@unito.it

Received: 21 September 2020; Accepted: 16 October 2020; Published: 21 October 2020

**Abstract:** The most important environmental characteristic in streams is flow. Due to the force of water current, most ecological processes and taxonomic richness in streams mainly occur in the riverbed. Benthic algae (mainly diatoms) and benthic macroinvertebrates (mainly aquatic insects) are among the most important groups in running water biodiversity, but relatively few studies have investigated their complex relationships. Here, we review the multifaceted interactions between these two important groups of lotic organisms. As the consumption of benthic algae, especially diatoms, was one of the earliest and most common trophic habits among aquatic insects, they then had to adapt to the particular habitat occupied by the algae. The environmental needs of diatoms have morphologically and behaviorally shaped their scrapers, leading to impressive evolutionary convergences between even very distant groups. Other less evident interactions are represented by the importance of insects, both in preimaginal and adult stages, in diatom dispersion. In addition, the top-down control of diatoms by their grazers contributes to their spatial organization and functional composition within the periphyton. Indeed, relationships between aquatic insects and diatoms are an important topic of study, scarcely investigated, the onset of which, hundreds of millions of years ago, has profoundly influenced the evolution of stream biological communities.

**Keywords:** coevolution; epizoosis; grazing; periphyton; scrapers

---

## 1. Introduction

"The substratum is the stage upon which the drama of aquatic insect ecology is acted out": this famous sentence of Minshall [1] is still relevant today, and can be extended to the essence of running water biota. In fact, while in lentic environments the water column hosts a rich and diverse community of producers (i.e., phytoplankton) and consumers (i.e., zooplankton and larger organisms), the effect of flow almost eliminates this habitat in lotic systems. Flow is the environmental factor that conditions life in lotic systems the most, exerting an incessant and unidirectional force on everything exposed to moving water [2]. Therefore, the majority of organisms that inhabit running water habitats belong to the benthic community, i.e., they are strictly associated with the stream bed. In this context, benthic algae and benthic macroinvertebrates represent the most important elements, but there are very few comprehensive studies that consider the different relationships between these two groups.

Almost all surfaces that receive light on river bottoms are covered with aquatic algae, which can be separated into two major non-taxonomic groups, macro- and micro-algae. The first group includes species whose thallus is visible to the naked eye, while the second group includes the smaller species. The latter is the most diverse and rich and, in this context, diatoms represent the ecologically most relevant group, reflecting their widespread occurrence and ecological success [3]. Diatoms are the most diverse group of algae in fresh waters and can colonize almost all aquatic habitats, although green algae and cyanobacteria can be very abundant under certain conditions. Filamentous green algae, for example, tend to prevail in lentic or slow water environments stressed by eutrophication, acidification and metal contamination [3]. The unique life cycle that characterizes diatoms, coupled with their particular wall structure and cell division mechanism [4], has probably played an important role in the evolutionary success of this group of algae. In the benthic habitats, primary producers are the principal component of the complex matrix called periphyton. Periphyton stoichiometry, i.e., C:N:P ratios, greatly influences grazer growth rate and the acquisition of energy reserves. Indeed, benthic algae represent a great energy source and high-quality food for macroinvertebrates due to their low C:N and C:P ratios and high protein and lipid contents [5]. Among the autotrophic groups composing the periphyton, diatoms are considered high quality food for macroinvertebrates compared to green algae [6] and cyanobacteria, due to their high content in long chain polyunsaturated fatty acids [7]. Beside stoichiometry, cyanobacteria represent a low-quality food source for grazers for several other reasons, such as the possession of toxins and secondary metabolites (for example protease inhibitors), their morphology (filamentous and/or colonial) and the lack of essential dietary lipids (see [8] and references therein).

Periphyton can be organized into different layers, whose spatial architecture approximates the vertical organization of terrestrial forests at a microscopic scale. In this context, diatoms are classified into growth forms according to their different positions in the biofilm, which in turn depend on their attachment mode to the substrate. At the basal level, we find adnate forms closely attached to the substrate through the entire valve surface or girdle bands. In the intermediate level, pad attached and stalked forms prevail. They first stick to the substrate by the mucilage excreted through the apical pore field located at one or both poles; the stalked forms produce a mucilage stalk that can be simple or branched. Within the periphyton, another important group of diatoms is represented by motile taxa; these are species able to move both within the biofilm and among cobbles towards the most suitable environmental conditions. Finally, on the upper layer we find overstory diatoms loosely attached, or unattached to the substrate. The growth form determines much of the diatom ecology, including how they interface with the primary consumers, discussed below.

Benthic macroinvertebrates represent a well-known group, widely used in biomonitoring, whose communities consist of species mainly belonging to Tricladida, Annelida, Crustacea, Mollusca, Nematoda, Nematomorpha, and Insecta. Insects represent the dominant component in lotic invertebrate communities because of their species richness, diversity, abundance, distribution and ubiquity [9]; 60.4% of the 126,000 worldwide known freshwater animal species are insects [10]. Likewise, aquatic insects are one of the most important elements of the lotic food webs, acting as primary and secondary consumers, pivotal in the processing and cycling of nutrients, producing a considerable amount of biomass to the ecosystem by means of their high secondary production, i.e., generating biomass [11,12]. They have also an important role linking aquatic food webs with terrestrial food webs by means of their flying activity during the adult stage of life.

Considering that diatoms and insects represent the most taxonomically rich, diverse and abundant groups among freshwater benthic algae and invertebrates, we focused on these two groups. Here, we review the complex interactions between the two most important groups of lotic organisms, namely benthic diatoms and macroinvertebrates, to try to shed light on the importance of these interactions.

The most obvious interaction between stream macroinvertebrates and diatoms is grazing, performed by herbivorous invertebrates on benthic biofilm. According to the classification of macroinvertebrates in functional feeding groups (FFGs; Table 1), the grazers (also called scrapers) include macroinvertebrates that feed on periphyton on hard surfaces [13].

**Table 1.** Classification of the functional feeding groups (FFGs).

| FFG | Type of Trophic Resources |
| --- | --- |
| Shredders | Coarse (>1 mm) particulate organic matter (CPOM): fragments of leaves, plant tissue and wood debris |
| Scrapers | Attached benthic algae (i.e., diatoms, green algae, and cyanobacteria) |
| Collector-gatherers | Fine (50 µm –1 mm) particulate organic matter (FPOM) |
| Collector-filterers | Very fine (<50 µm) particulate organic matter and micro-organisms in the water column |
| Predators | Small animals |

Although many stream invertebrates have considerable plasticity in the foods they ingest and can consume periphyton at some part of their lives [14], we found specialized scrapers in 6 out of 13 orders of insects with aquatic species not particularly related among them from a phylogenetic point of view. This shows that this trophic strategy has evolved several times, independently, in running water environments.

For those groups that colonized fresh waters originating from saltwater environments (e.g., gastropod molluscs, such as Lymnaeidae, Physidae) scraping is a simple transposition of something that occurred in marine ancestors. Nevertheless, the situation is different for aquatic insects, the dominant group of river macroinvertebrates, in terms of both taxonomic [10] and trophic [15] diversity.

It is generally believed that the first insects evolved in terrestrial environments and subsequently colonized inland waters, although some discordant hypotheses have emerged [16]. The first aquatic insects probably date back to the Permian [9]. Several groups of insects independently invaded aquatic habitats [17], so that truly aquatic insects include the most primitive winged forms (Ephemeroptera) and other ancient Paleoptera (Odonata), but also Neoptera Exopterygota (Plecoptera) and Endopterygota (Trichoptera and Megaloptera). In addition to this, typically aquatic taxa are present in about eight other insect orders, such as Coleoptera, Diptera, Lepidoptera and Hemiptera [9]. Most of these groups comprise scrapers. Moreover, even some taxa usually considered as predators can behave as mainly diatom consumers in some particular habitats, such as temporary streams [18].

In this context, two elements emerge as underlining the strict and ancient relationship between aquatic insects and benthic algae.

The first element is antiquity. The first aquatic insects are generally believed to have been predators feeding on a broad range of invertebrates. The second trophic habit that appeared was benthic periphyton-scraping, in the middle of the Permian period [19]. This hypothesis comes from the analysis of trophic resources available in ancient lotic systems (see [20]). Firstly, we have no fossil traces of aquatic macrophytes before the Triassic; moreover, macrophytes do not represent a primary food source for benthic invertebrates and, in particular, for insects [21]. Furthermore, it is well known that an important part of the energy input in current lotic systems derives from allochthonous non-living coarse particulate organic matter (CPOM), mainly terrestrial leaves [22]. However, this resource, consumed by shredders, was probably scarcely available in the past because ancient catchments were barely vegetated, and CPOM only assumed a certain importance in the early Cretaceous period, with the spread of angiosperms [20]. Consequently, the availability of fine organic particles (FPOM), originating from CPOM breakdown, was also probably low, explaining the late advent of filterers [23]. It can therefore be hypothesized that, for a very long period, the direct consumption of periphyton was one of the earliest, most important and common trophic habits among lotic invertebrates, and in particular among insects.

The second element is the selective pressure. As mentioned before, a second interesting aspect underlining the important relationship between benthic algae and aquatic insects is that scraping appeared several times, independently, in different and even very phylogenetically distant insect groups. Scrapers can be found in six orders of insects and for some of these, feeding on algae represents probably one of the most important ways of life [23]. In the most ancient winged group, the Ephemeroptera, grazers are very common in many families, such as Heptageniidae, Siphlonuridae, Leptophlebiidae, Caenidae and Baetidae [24]. In addition, many Plecoptera feed on periphyton, such as some Diamphipnoidae, Gripopterygidae, Notonemouridae, Capniidae and Taeniopterygidae [25,26]. Among Hemiptera, some Corixidae are largely herbivorous, scraping algae from submerged vegetation and stone [9,27]. Among Endopterygota, scrapers are also very common, for example, Coleoptera Hydraenidae, Psephenidae, Helodidae/Scirtidae, Dryopidae and Elmidae feed on algae in the preimaginal and/or adult stage [13]. Several families of Trichoptera are almost exclusively scrapers, such as Glossosomatidae, Helicopsychidae, Apataniidae, Goeridae and Uenoidae; others include many taxa with this trophic habit, such as Hydroptilidae and Leptoceridae, while some other taxa are facultative scrapers, such as Psychomyiidae, *Hydropsyche* (Hydropsychidae) and *Marila* (Odontoceridae) [28]. Among Diptera, according to [29], larval Blephariceridae, Deuterophlebiidae, some Psychodidae, Ephydridae and Chironomidae are scrapers. In practice, all these organisms, even very distant from a phylogenetic point of view, have undergone the same evolutionary pressures to adapt to the scraper diet, and have therefore evolved some similar characteristics in a convergent way.

What are these characteristics? Or rather, how did algae shape their consumers? We must first consider that the development of periphyton is not homogeneous in all river microenvironments, being favored by some specific environmental conditions, such as good sunlight exposure (which facilitates the photosynthetic process), medium to high current velocity (which prevents sedimentation and thus protects algae from burial) and, in general, coarse substrata (which ensures better immovability over time and a more stable colonization) [30].

In scrapers belonging to different insect groups, natural selection has favored the development of similar morphological and behavioral adaptations, which can be summarized as follows.

## 2. Ability to Withstand Elevated Velocity Current Environments

Compared to shredders or predators, lotic grazers are exposed to higher hydraulic stress, as their food sources colonize the top of boulders and pebbles due to their light requirements. Therefore, the general adaptations presented by lotic aquatic insects to the high flow speed must be more developed and more sophisticated in scrapers.

Different convergent strategies have been developed, aimed to minimize the threat of drifting downstream by the current. These essentially include adopting a hydrodynamic profile, having powerful adhesion structures (nails, suction cups, silk) or, in some cases, weighing down the body. Among aquatic insects, the most impressive examples of hydrodynamic shape are found in scrapers. Since the pioneering studies of Statzner and collaborators [31], the dorsoventral flatness of Heptageniidae (Ephemeroptera) is believed to be the key that enabled these animals to colonize even the fastest environments, taking shelter in the lowest, thick and viscous portion of the boundary layer, where flow is slowest and laminar (Figure 1). In addition, immature stages of Diptera Deuterophlebiidae, typical scrapers that inhabit riffle habitats where current velocities usually exceed 1 m/s, are noticeably dorsoventrally flattened [32]. The same adaptation has occurred several times in the evolutionary history of aquatic insects, reaching the most spectacular forms among Coleoptera, where the larvae of Elmidae and Psephenidae have unique dimensional relationships, with very flattened, shield- to disc-shaped body forms [17]. In addition to dorsoventral flattening, mayflies and beetles, despite being distant in the evolutionary tree of hexapods, have adopted the same solutions even in smaller details, namely the general body form (broader at the front and narrow behind), the presence of smooth lateral structures helping the hydrodynamic profile (i.e., femura and headparts in heptagenids, lateral plate-like extensions in Coleoptera larvae) and the existence of marginal hair fringes that increase the adhesion capacity to the substrate.

**Figure 1.** Dorsoventral flattening of an *Ecdyonurus* (Ephemeroptera: Heptageniidae) nymph (Photo Roberto Messori; reproduced with permission from the author).

Moreover, to live in fast-flowing zones with high hydraulic constraints, scrapers have developed very effective adhesion structures. The tarsal claws of rheophilic Ephemeroptera, Plecoptera, Coleoptera, Diptera and Trichoptera are usually extremely robust, sharp and curved to better interlock with the surface irregularities of the substrate [17]. Some Heptageniidae, such as *Epeorus assimilis*, can cling with their claws to even slightly rough substrates, but Ditsche-Kuru and collaborators [33] found that the presence of biofilm considerably increases the adhesion ability of insects. Other attachment devices are represented by circlets of small hooks in the pseudopoda of Deuterophlebiidae and powerful ventral suckers of Blephariceridae (Figure 2).

**Figure 2.** Ventral suckers of a Blephariceriidae larva (Photo Stefano Fenoglio).

## 3. Ability to Scrape Algae from the Substrate

Lotic algae, and in particular diatoms, are known to have strong adhesion ability to hard substrates [34,35]. To feed on these organisms, different insect groups have had to adopt similar morphological adaptations, in particular regarding mouthpart morphology. Arens [36] reported that all scrapers had to find a solution to solve four main problems. Firstly, they must be able to detach the algae from the substrate. Secondly, once removed from the substrate, algae must be collected and transferred to the oral cavity. For example, some Heptageniidae mayflies exhibit adapted labial and maxillary palps combining reinforced scraping parts and setose brush parts [37]. Among Plecoptera, Brachypterinae have maxillae combining chisel-like structures in the lacinia and brush-like structures in the galea [17]. Among Trichoptera, the mandibles of Glossosomatidae are elongated with edge and broom structures [28]. In scraping beetles, mandibles can present a sharp, anterior incisor lobe, while galea or lacinia can show a brush apparatus. Thirdly, the diatom siliceous shell can be a mechanical protection against grazing that grazers need to overcome, although its defensive effectiveness is not supported by much experimental evidence [38]. Finally, all this mouthpart activity must be carried out in the shelter of the current. It is impressive how, starting from different morphological bases, all insect scrapers have converged in adopting very similar solutions; for more details, see the comparison made by Arens [39] of scraper mouthparts using SEM images. Lastly, this feeding mode leads to another interesting convergence among aquatic insects, as scraper mouthparts are subjected to strong abrasion and wear out quickly. These organisms have thus had to evolve specific adaptations, or use some previously existing ones, such as the exoskeleton moult in arthropods, to favor renewal of mouthparts and prolong their duration (see details in [36]).

In reviewing the aquatic insects–periphyton relationships, Lamberti and Moore [40] wondered if taxon-specific preferences between grazers and attached algae exist and to what extent this selectivity can be considered as an active or passive selection. After decades of field and mesocosm experiments, the scientific literature shows evidence for a certain degree of selectivity by scrapers but, generally, freshwater ecologists agree upon the fact that this selectivity is mainly a passive consequence of the interplay between the size and morphology of the insect mouthparts and the algal susceptibility to grazing according to their life forms [41–43]. Indeed, as already mentioned before, pad attached (such as *Meridion*) or stalked (such as *Cymbella* and *Gomphonema*) diatoms are more susceptible to grazing pressure than small adnate forms (such as *Achnanhtidium*) [41].

## 4. Adaptation to Live in The Open

Most benthic invertebrate species live hidden among the elements of the river bottom, between the detritus, under the large boulders, in the interstices between the pebbles, or even burrowed in the sand, where they find food and shelter from predators. However, this strategy cannot be adopted by scrapers, which must spend a lot of time "above" and not "below" the rocks to feed. For this reason, when wading a stream, insect scrapers are the most easily observable: for example, in an alpine creek, the nymphs of Heptageniidae are commonly visible on the upper surface of boulders, while dense populations of Blephariceridae stand out on large rocks with fast flow. It is the very nature of their microhabitat (and therefore their food strategy) that protects them from predators: in fact, predators select the most advantageous prey according to different factors, such as prey density, energy contents, handling time and encounter rate [44], and this last element greatly reduces the vulnerability of scrapers. For example, analyzing the diet of two large-sized predaceous Systellognatha Plecoptera, Bo et al. [45] found that Blephariceridae, even if present in the same river stretch, were never consumed, probably because they were too difficult to reach. Moreover, in another study focused on feeding preferences of predaceous stoneflies, using trophic electivity indices to compare gut contents with natural composition and abundance of macroinvertebrate community in the riverbed, Heptageniidae, as a food source, was always negatively selected [46]. Interestingly, we also report that the few non-scraper insects inhabiting the same microhabitats (upper surface of masses and rocks in very fast-flowing waters), such as Diptera Simuliidae, are only included in the predator diet in exceptional cases [46].

## 5. Direct and Indirect Effects of Scrapers in Shaping Diatom Communities

The relationship between scrapers and diatoms can also be viewed from the other side: how do scrapers shape the diatom community as an effect of herbivory? Did diatoms evolve defense mechanisms against herbivory? Grazing in freshwater benthic ecosystems was essentially unexplored until the 1980s when Gregory [47] elaborated the theory of a top-down control of primary producers by grazers. In the book Algal Ecology, Stevenson et al. [48] devoted an entire chapter to the regulative action on benthic algae played by grazers. In most cases, periphyton is considered to be affected by both bottom-up (nutrients, light and other abiotic factors) and top-down controls [49], but their respective role can vary upon additional conditions. Graça and collaborators [49] performed an experimental study in a low order tropical stream in enclosures under controlled conditions of nutrients and light. Given the low grazer density, they expected a low grazing pressure. Conversely, they found an intense grazing effect, more evident in semi-shaded areas (53% algal biomass removed) than in fully shaded ones (33%). Moreover, grazing resulted in a decrease of diatoms in favor of cyanobacteria, which are considered more tightly adhered in some cases (e.g., Chamaesiphonales), less palatable and lacking in high-saturated fatty acids.

Herbivores can frequently alter periphyton composition and physiognomy as they more easily consume overstory or loosely-attached algae [50], favoring a shift towards tightly attached prostrate forms. Despite this, a recent study highlighted that herbivorous abundance apparently does not significantly affect diatom ecological guild structure, and that grazing pressure has no effects on diatom ecological guild diversity [51]. Besides possible changes in community composition, the overall biomass can be almost unaltered by grazing pressure. In fact, the ability of grazers to induce a significant biomass loss depends, among other reasons, on their density, as shown in classical mesocosm studies [52].

Given the complex multilayer structure of periphyton and the different feeding strategies adopted by grazers, it is reasonable to hypothesize preferential feeding based on resource partitioning. Resource partitioning is a central concept in community ecology as it may explain the coexistence of species belonging to the same guild. It has been confirmed for terrestrial vertebrates, especially birds [53] but hardly associated with benthic grazers, which are generally considered as largely generalists. In a recent study, Piano et al. [54] examined the distribution of three Heptagenidae taxa, namely *Rhitrogena*, *Epeorus* and *Ecdyonurus*, commonly found in Alpine streams. Their findings suggested that the distribution of the first two genera was strictly linked to diatom biomass, whereas this relationship was not so evident for *Ecdyonurus*, which has a more generalist diet including allochthonous detritus. Tall et al. [55] analyzed the food preferences of grazers feeding on epiphytic diatoms growing on the moss *Fontinalis*. Their findings suggest that when available food is reduced there is a resource partitioning within the grazers. Based on their results, they divided the grazers into three categories: (1) true scrapers, such as the coleopteran *Promoresia* and the caddisfly *Hydroptila*, which fed selectively on the adnate diatoms; (2) generalists (e.g., *Baetis* and some Chironomidae), without any clear preference regarding the biofilm layer; (3) surfers (e.g., some Chironomidae), which avoided the adnate diatoms and showed a preference for detached cells. In this repartition, mouthpart morphology has a great importance [56].

In addition to the direct effects of grazing, aquatic insects that feed on attached algae can also affect the composition and biomass of periphyton indirectly. For instance, the action of the grazers may reduce the self-shading of biofilms and, in turn, stimulate the cell growth and species turnover within the periphytic layer [57]. Moreover, grazers can indirectly affect diatom composition by favoring the dispersal of certain taxa. Indeed, Peterson [41] observed that after gut passage, an important percentage of ingested diatom cells stay viable when eliminated through feces (about 40–52%). The physical dislodgement of diatoms by grazers can, therefore, significantly contribute to diatom dispersal, especially in the downstream stretches. Another indirect and beneficial effect attributed to grazers is the nutrient enrichment due to the residuals of ungrazed algae as well as grazers' feces. In a mesocosm experiment, Herren et al. [58] found that the area-specific primary production of attached algae was 71% higher in the presence of residential Chironomidae larvae than in their absence, probably because of the consumers' fertilizing action. These findings suggest that the relationship between insect scrapers and attached algae is a complex trade-off between positive (i.e., beneficial) and negative (i.e., adverse) interactions [57].

The occupation of different layers within the periphyton and, in particular, the basal one can be seen as a strategy to resist the action of grazing. Furthermore, the siliceous frustule could have a defensive function, though its anti-grazing role has been questioned in some articles, almost all related to marine planktonic environment. Hamm et al. [59] started from the consideration that its role is not decisive as diatoms are an elective food for a large variety of organisms. Nevertheless, they highlighted that in an environment characterized by a high grazing pressure, such as the planktonic one, the presence of a mechanical protection should play a role in reducing diatom population mortality and in shaping phyto- and zoo-plankton populations. Pančić et al. [38] experimentally demonstrated that the frustule defensive power of marine phytoplanktonic diatoms varies according to different grazing modes. Calanoid copepods have robust mandibles that can crack diatom frustules before ingestion, with a success rate that depends on the silica content of the frustule, which can vary among diatom species. The frustule does not provide any protection against grazing by protozoans, which engulf their prey and digest the cell content without breaking the valves. Despite the paucity of similar studies in freshwater benthic environments, we can confirm that aquatic insects often digest the diatom content without destroying the frustules. In fact, complete diatom frustules are usually found in aquatic insect gut contents, and even intact but empty frustules are found in the insect fecal material, together with other diatom individuals that have survived the passage through the digestive system [18,60].

Producing harmful secondary metabolites as anti-predatory mechanisms is a complex and interesting issue that must still be unraveled. A few diatom taxa are known to produce a toxic amino acid, domoic acid, which causes serious gastrointestinal and neurological consequences in humans and aquatic animals [61]. These are the marine genus *Pseudonitzschia* [61] and the estuarine species *Amphora coffeaeformis* [62]. More recently, Violi et al. [63] reported the production of other toxic amino acids, e.g., β-methylamino-L-alanine and its isomer 2,4-diaminobutyric acid, by several freshwater diatoms under certain culture conditions, formerly only attributed to Cyanobacteria. They hypothesized that diatoms could produce these toxins as a response to stress such as nutrient depletion. As suggested by this study, production of these toxins in freshwater environments may become a health issue in freshwater diatom blooms, although the effect on aquatic insects requires further research to be assessed.

On the other hand, the effects that massive blooms of some benthic diatoms have in modifying the habitat for river communities are quite well known. In this regard, an emblematic example is represented by the blooms of *Didymosphenia geminata*. This is the most studied invasive diatom [64], which under certain environmental conditions produces large quantities of extracellular stalks that can almost completely cover more or less long stretches of river, sometimes reaching an extension of a few kilometers in length [65]. The blooms of *D. geminata* can profoundly alter the invaded benthic communities, by decreasing β-diversity and increasing taxonomic homogenization in both algal and invertebrate assemblages [66]. The thick filament mats prevent the movements of the larger aquatic insects, such as the Heptageniidae, favoring instead smaller, opportunistic and generally herbivorous organisms, such as the chironomids and oligochaetes [65,66]. When in bloom, the *D. geminata* filamentous mats cover almost completely hard substrates, disadvantaging crawlers, shredders and scrapers. The chironomids take advantage of the absence of large competitors and predators, and a significant change in top-down control is therefore determined because the chironomids prefer the larger diatoms, favoring in turn the smaller and pioneering taxa. Even though *D. geminata* stalks can provide a suitable substrate for other diatoms (such as *Achnanthidium* spp.), contributing to changes in diatom community composition, we found no significant evidence that this could lead to advantages or disadvantages on macroinvertebrate communities. Overall, *D. geminata* blooms lead to a simplification of the food web structure with a dominance of smaller organisms.

## 6. Epizoic Relationships

In some instances diatoms take advantage of the relationship with macroinvertebrates, as in the case of the epibiosis. The term "epibiosis" describes a strict association between organisms, in which we can identify an epibiont (the organism that lives attached to the body surface of another organism) and a basibiont (the organism that hosts epibionts [67]). Even if is not parasitic, this relationship is often unbalanced because the epibiont has the greatest advantages. Here we review the relationship between diatoms (as epibionts) and aquatic insects (as basibionts) in running water ecosystems.

Epibionts, such as benthic freshwater diatoms, have some obvious advantages in colonizing the surfaces of aquatic insects, including: (i) an enhanced protection against grazing; (ii) an increased accessibility to solar radiation, carbon dioxide and nutrients, because of the basibiont activity and metabolism; (iii) a better ability to avoid burial by fine sediments; and (iv) an important benefit for dispersion [68]. Benefits for basibionts are less obvious; some studies report that the presence of an epibiont biofilm can be useful to increase camouflage and reduce irradiation, while others emphasize how this layer can impair numerous functional processes (such as gas exchange) and limit the possibility of movement [69]. For these reasons, epizoic (i.e., living on an animal) diatoms are probably more common and diverse than previously supposed. To date, most studies on epizoic diatoms have been performed in marine environments [70] and, of the studies carried out in fresh water, many have focused on non-insect taxa. Freshwater diatoms have been found on Testudines [71], Gastropoda Physidae [72], Rotifera [73], Copepoda [74] and, in particular, Crustacea Decapoda. Due to their generally large size and hard exoskeletons, freshwater crabs and crayfish ideally represent optimal basibionts for diatoms and other microalgae [68,75].

Although there are not many studies on the diatom–insect epibiosis, we can still underline how this relationship has a considerable and probably underestimated importance, especially regarding the role of aquatic insects in algal dispersion. Very few studies have focused on diatoms living on preimaginal stages of aquatic insects. Larval cases of some Trichoptera are known to host algae [76], but mechanisms regulating epibiotic associations between diatoms and aquatic stages are still practically unknown; not all aquatic insects nor all diatom species show this kind of association. For example, Wujek [77], using scanning electron microscopy, reported that three species of diatoms, among the numerous present in the substrate of Cedar Creek (MI, USA), lived on the cerci of Caenidae nymphs, while none were found on the nymphs of sympatric Ephemerellidae. Further studies need to be carried out regarding pre-imaginal stages, while we have more information about epizoic diatoms on adult, winged organisms. In fact, most aquatic insects have an "amphibious" life-cycle, characterized by the presence of a pre-imaginal aquatic stage and an imaginal terrestrial stage. Contact between diatoms and adult insects may occur during emergence from nymphal or pupal exuvia, or during oviposition, when females (and occasionally also males, e.g., in some Odonata) come into contact with surface water or even submerge to lay eggs in the vegetation or substrate, or when resting in the wet areas on the banks of the stream or on the boulders reached by the splashes of water. For example, Stewart and Schlichting [78] reported the presence of diatoms on the exoskeleton of some Odonata Zygoptera and Anisoptera, Hemiptera Gelostocoridae, Diptera Chironomidae and Trichoptera Polycentropodidae collected with light-traps at night or picked from riverbanks and vegetation surrounding water. Another study reported that some diatoms (such as *Navicula* and *Nitzschia*) were found attached to adult aquatic Diptera belonging to Tipulidae and Ptychopteridae families [79].

This is interesting, because adult stages may represent an important element in the dispersion of diatoms, allowing algae to colonize new environments and to pass from river to river across land barriers.

## 7. Conclusions

In conclusion, relationships between aquatic insects and diatoms is a subject of great interest, which, over hundreds of millions of years, has profoundly marked the evolutionary path of biological communities in streams. In running water environments, these groups are truly among the most important organisms from different points of view, whose multiple and often still unknown relationships form the substrate at the base of entire communities and trophic chains. This review shows that the direct and indirect grazing behavior of aquatic insects acts as an important and multifaceted mechanism that affects the diversity, composition, growth rate and biomass of attached algae, along with the bottom-up processes. However, the effects of these biotic interactions in shaping benthic algae communities and ecosystem functionality have not been studied in-depth (but see [80,81]). A novel and challenging approach could be adopting the recent advances of the metacommunity theory [82], as biological communities are simultaneously shaped by the pure and combined effect of environmental filters, spatial variables and biotic interactions. Although the former two categories of predictors have been investigated a lot, few studies have attempted to quantify the latter category. In light of their strong evolutive and trophic relationship, we believe that attached algae and aquatic insect scrapers represent ideal organisms to better investigate the role of biotic interactions in shaping benthic metacommunities. Similarly, we believe that all the natural and human-induced variations in the distribution and density of scrapers, such as flow and thermal and sedimentary alterations, deserve greater attention for future studies because of their cascade effects on the attached algae.

**Author Contributions:** All authors contributed equally to this paper. All authors have read and agreed to the published version of the manuscript.

**Funding:** This research received no external funding.

**Acknowledgments:** This work was realized within the framework of the activities of ALPSTREAM, a research center financed by FESR, Interreg Alcotra 2014–2020, EcO Project of the Piter Terres Monviso. English language review supplied by Rhadika Srinivasan.

**Conflicts of Interest:** The authors declare no conflict of interest.

## References

1. Minshall, G.W. Aquatic insect-substratum relationships. In *The Ecology of Aquatic Insects*; Resh, V.H., Rosenberg, D.M., Eds.; Praeger Publishers: New York, NY, USA, 1984; pp. 358–400.
2. Poff, N.L.; Allan, J.D.; Bain, M.B.; Karr, J.R.; Prestegaard, K.L.; Richter, B.D.; Sparks, R.E.; Stromberg, J.C. The natural flow regime. *BioScience* **1997**, *47*, 769–784. [CrossRef]
3. Bellinger, E.G.; Sigee, D.C. *Freshwater Algae: Identification, Enumeration and Use as Bioindicators*; John Wiley Sons: London, UK, 2015; p. 290.
4. Chepurnov, V.A.; Mann, D.G.; Sabbe., K.; Vyverman, W. Experimental studies on sexual reproduction in diatoms. *Int. Rev. Cytol.* **2004**, *237*, 91–154. [PubMed]
5. Torres-Ruiz M., J.D.; Wehr, J.D.; Perrone, A.A. Trophic relations in a stream food web: Importance of fatty acids for macroinvertebrate consumers. *J. North. Am. Benthol. Soc.* **2007**, *26*, 509–522. [CrossRef]
6. Guo, F.; Kainz, M.J.; Sheldon, F.; Bunn, S.E. The importance of high-quality algal food sources in stream food webs–current status and future perspectives. *Freshw. Biol.* **2016**, *61*, 815–831. [CrossRef]
7. Guo, F.; Bunn, S.E.; Brett, M.T.; Fry, B.; Hager, H.; Ouyang, X.; Kainz, M.J. Feeding strategies for the acquisition of high-quality food sources in stream macroinvertebrates: Collecting, integrating, and mixed feeding. *Limnol. Oceanogr.* **2018**, *63*, 1964–1978. [CrossRef]
8. Groendahl, S.; Fink, P. High dietary quality of non-toxic cyanobacteria for a benthic grazer and its implications for the control of cyanobacterial biofilms. *BMC Ecol.* **2017**, *17*, 20. [CrossRef]
9. Merritt, R.W.; Wallace, J.B. Aquatic habitats. In *Encyclopedia of Insects*; Resh, V.H., Cardi, R.T., Eds.; Academic Press: Burlington, MA, USA, 2009; pp. 38–48.
10. Balian, E.V.; Segers, H.; Martens, K.; Lévéque, C. The freshwater animal diversity assessment: An overview of the results. *Hydrobiologia* **2008**, *595*, 627–637. [CrossRef]

11. Resh, V.H.; Rosenberg, D.M. *The Ecology of Aquatic Insects*; Praeger Publisher: New York, NY, USA, 1984; p. 625.

12. Cummins, K.W.; Merritt, R.W. Ecology and distribution of aquatic insects. In *An Introduction to the Aquatic Insects of North. America*; Merritt, R.W., Cummins, K.W., Eds.; Kendall Hunt Publishing Company: Dubuque, IA, USA, 1996; pp. 74–86.

13. Cummins, K.W. Trophic relations of aquatic insects. *Annu. Rev. Entomol.* **1973**, *18*, 183–206. [CrossRef]

14. Hauer, F.R.; Lamberti, G.A. *Methods in Stream Ecology*; Academic Press: New York, NY, USA, 2011; p. 506.

15. Fenoglio, S.; Merritt., R.W.; Cummins, K.W. Why do no specialized necrophagous species exist among aquatic insects? *Freshw. Sci.* **2014**, *33*, 711–715. [CrossRef]

16. Pritchard, G.; McKee, M.H.; Pike, E.M.; Scrimgeour, G.J.; Zloty, J. Did the first insects live in water or in air? *Biol. J. Linn. Soc.* **1993**, *49*, 31–44. [CrossRef]

17. Wichard, W.; Arens, W.; Eisenbeis, G. *Biological Atlas of Aquatic Insects*; Apollo Books: Aamosen, Denmark, 2002; p. 339.

18. López-Rodríguez, M.J.; Tierno de Figueroa, J.M.; Fenoglio, S.; Bo, T.; Alba-Tercedor, J. Life strategies of 3 Perlodidae species (Plecoptera) in a Mediterranean seasonal stream in southern Europe. *J. North. Am. Benthol. Soc.* **2009**, *28*, 611–625. [CrossRef]

19. Wooton, R.J. The historical ecology of aquatic insects: An overview. *Palaeogeogr. Palaeoclimatol. Palaeoecol.* **1988**, *62*, 477–492. [CrossRef]

20. Lancaster, J.; Downes, B.J. *Aquatic Entomology*; Oxford University Press: Oxford, UK, 2013; p. 296.

21. Allan, J.D.; Castillo, M.M. *Stream Ecology: Structure and Function of Running Waters*, 2nd ed.; Springer: Dordrecht, The Netherlands, 2007.

22. Vannote, R.L.; Minshall, G.W.; Cummins, K.W.; Sedell, J.R.; Cushing, C.E. The river continuum concept. *Can. J. Fish. Aquat. Sci.* **1980**, *37*, 130–137. [CrossRef]

23. Huryn, A.D. Aquatic insects—Ecology, feeding, and life history. In *Encyclopedia of Inland Waters*; Likens, G., Benbow, M.E., Burton, T.M., Van Donk, E., Downing, J.A., Gulati, R.D., Eds.; Academic Press: New York, NY, USA, 2009; pp. 25–50.

24. Brittain, J.E.; Sartori, M. Ephemeroptera. In *Encyclopedia of Insects*; Resh, V.H., Cardé, R.T., Eds.; Academic Press: Burlington, MA, USA, 2009; pp. 328–333.

25. Fenoglio, S.; Bo, T.; López-Rodríguez, M.J.; Tierno de Figueroa, J.M. Nymphal biology of *Brachyptera risi* (Morton, 1896) (*Plecoptera: Taeniopterygidae*) in a North Apennine stream (Italy). *Entomol. Fenn.* **2008**, *19*, 228–231. [CrossRef]

26. Tierno de Figueroa, J.M.; López-Rodríguez, M.J. Trophic ecology of *Plecoptera* (Insecta): A review. *Eur. Zool. J.* **2019**, *86*, 79–102. [CrossRef]

27. Haedicke, C.W.; Redei, D.; Kment, P. The diversity of feeding habits recorded for water boatmen (*Heteroptera: Corixoidea*) world-wide with implications for evaluating information on the diet of aquatic insects. *Eur. J. Entomol.* **2017**, *114*, 147–159. [CrossRef]

28. Holzenthal, R.W.; Thomson, R.E.; Ríos-Touma, B. Order Trichoptera. In *Thorp and Covich's Freshwater Invertebrates: Ecology and General Biology*; Thorp, J.H., Rogers, D.C., Eds.; Elsevier: London, UK, 2015; Volume 1, pp. 965–1002.

29. Courtney, G.W.; Cranston, P.S. Diptera. In *Thorp and Covich's Freshwater Invertebrates: Ecology and General Biology*; Thorp, J.H., Rogers, D.C., Eds.; Elsevier: London, UK, 2015; Volume 1, pp. 1043–1058.

30. Biggs, B.J.F. Patterns in benthic algae of streams. In *Algal Ecology: Freshwater Benthic Ecosystems*; Stevenson, R.J., Bothwell, M.L., Lowe, R.L., Eds.; Academic Press: San Diego, CA, USA, 1996; pp. 31–56.

31. Statzner, B.; Holm, T.F. Morphological adaptations of benthic invertebrates to stream flow—An old question studied by means of a new technique (Laser Doppler Anemometry). *Oecologia* **1982**, *53*, 290–292. [CrossRef]

32. Courtney, G.W. Life history patterns of Nearctic mountain midges (*Diptera: Deuterophlebiidae*). *J. North. Am. Benthol. Soc.* **1991**, *10*, 177–197. [CrossRef]

33. Ditsche-Kuru, P.; Barthlott, W.; Koop, J.H. At which surface roughness do claws cling? Investigations with larvae of the running water mayfly *Epeorus assimilis* (*Heptageniidae, Ephemeroptera*). *Zoology* **2012**, *115*, 379–388. [CrossRef]

34. Higgins, M.J.; Crawford, S.A.; Mulvaney, P.; Wetherbee, R. Characterization of the adhesive mucilages secreted by live diatom cells using atomic force microscopy. *Protist* **2002**, *153*, 25–38. [CrossRef]

35. Gebeshuber, I.C.; Thompson, J.B.; Del Amo, Y.; Stachelberger, H.; Kindt, J.H. In vivo nanoscale atomic force microscopy investigation of diatom adhesion properties. *Mater. Sci. Tech. Lond.* **2002**, *18*, 763–766. [CrossRef]

36. Arens, W. Wear and tear of mouthparts: A critical problem in stream animals feeding on epilithic algae. *Can. J. Zool.* **1990**, *68*, 1896–1914. [CrossRef]

37. McShaffrey, D.; McCafferty, W.P. Feeding behavior of *Rhithrogena pellucida* (Ephemeroptera: Heptageniidae). *J. North. Am. Benthol. Soc.* **1988**, *7*, 87–99. [CrossRef]

38. Pančić, M.; Torres, R.R.; Almeda, R.; Kiørboe, T. Silicified cell walls as a defensive trait in diatoms. *Proc. R. Soc. B Biol. Sci.* **2019**, *286*, 20190184.

39. Arens, W. Striking convergence in the mouthpart evolution of stream-living algae grazers. *J. Zool. Syst. Evol. Res.* **1994**, *32*, 319–343. [CrossRef]

40. Lamberti, G.A.; Moore, J.W. Aquatic insects as primary consumers. In *The Ecology of Aquatic Insects*; Resh, V.H., Rosenberg, D.M., Eds.; Praeger: New Yok, NY, USA, 1984; pp. 164–195.

41. Peterson, C.G. Gut passage and insect grazer selectivity of lotic diatoms. *Freshwat. Biol.* **1987**, *18*, 455–460. [CrossRef]

42. Díaz Villanueva, V.; Albariño, R. *Algal Ingestion and Digestion by Two Ephemeropteran Larvae from a Patagonian Andean Stream*; Research update on Ephemeroptera & Plecoptera; University of Perugia: Perugia, Italy, 2003; pp. 468–475.

43. Schmid-Araya, J.M.; Figueroa Hernández, D.; Schmid, P.E.; Drouot, C. Algivory in food webs of three temperate Andean rivers. *Austral. Ecol.* **2012**, *37*, 440–451. [CrossRef]

44. Krebs, J.R. Optimal foraging: Decision rules for predators. In *Behavioural Ecology: An Evolutionary Approach*; Krebs, J.R., Davies, N.B., Eds.; Blackwell: Oxford, UK, 1978; pp. 23–63.

45. Bo, T.; Fenoglio, S.; López-Rodríguez, M.J.; Tierno de Figueroa, J.M. Trophic behavior of two Perlidae species (Insecta, *Plecoptera*) in a river in southern Spain. *Int. Rev. Hydrobiol.* **2008**, *93*, 167–174. [CrossRef]

46. Bo, T.; Fenoglio, S.; Malacarne, G. Diet of *Dinocras cephalotes* and *Perla marginata* (Plecoptera: Perlidae) in an Apennine stream (northwestern Italy). *Can. Entomol.* **2007**, *139*, 358–364. [CrossRef]

47. Gregory, S.V. Plant-herbivore interactions in stream systems. In *Stream Ecology. Application and Testing of General Ecological Theory*; Barnes, J.R., Minshall, G.W., Eds.; Springer: Boston, MA, USA, 1983; pp. 157–189.

48. Stevenson, R.J.; Bothwell, M.L.; Lowe, R.L.; Thorp, J.H. *Algal Ecology: Freshwater Benthic Ecosystem*; Academic press: Dubuque, MA, USA, 1996; p. 753.

49. Graça, M.A.; Callisto, M.; Barbosa, J.E.L.; Firmiano, K.R.; França, J.; Júnior, J.F.G. Top-down and bottom-up control of epilithic periphyton in a tropical stream. *Freshwat. Sci.* **2018**, *37*, 857–869. [CrossRef]

50. Holomuzki, J.R.; Feminella, J.W.; Power, M.E. Biotic interactions in freshwater benthic habitats. *J. North. Am. Benthol. Soc.* **2010**, *29*, 220–244. [CrossRef]

51. Vilmi, A.; Tolonen, K.T.; Karjalainen, S.M.; Heino, J. Metacommunity structuring in a highly-connected aquatic system: Effects of dispersal, abiotic environment and grazing pressure on microalgal guilds. *Hydrobiologia* **2017**, *790*, 125–140. [CrossRef]

52. Steinman, A.D.; McIntire, C.D.; Gregory, S.V.; Lamberti, G.A. Effects of irradiance and grazing on lotic algal assemblages. *J. Phycol.* **1989**, *25*, 478–485. [CrossRef]

53. MacArthur, R.H. Population ecology of some warblers of north eastern coniferous forests. *Ecology* **1958**, *39*, 599–619. [CrossRef]

54. Piano, E.; Doretto, A.; Falasco, E.; Gruppuso, L.; Fenoglio, S.; Bona, F. The role of recurrent dewatering events in shaping ecological niches of scrapers in intermittent Alpine streams. *Hydrobiologia* **2019**, *841*, 177–189. [CrossRef]

55. Tall, L.; Cattaneo, A.; Cloutier, L.; Dray, S.; Legendre, P. Resource partitioning in a grazer guild feeding on a multilayer diatom mat. *J. North. Am. Benthol. Soc.* **2006**, *25*, 800–810. [CrossRef]

56. Merritt, R.; Cummins, K.; Berg, M.N. *An Introduction to the Aquatic Insects of North. America, 5th ed*; Kendall Hunt: Dubuque, IA, USA, 2019; p. 498.

57. Vadeboncoeur, Y.; Power, M.E. Attached algae: The cryptic base of inverted trophic pyramids in freshwaters. *Annu. Rev. Ecol. Evol. S.* **2017**, *48*, 255–279. [CrossRef]

58. Herren, C.M.; Webert, K.C.; Drake, M.D.; Jake Vander Zanden, M.; Einarsson, Á.; Ives, A.R.; Gratton, C. Positive feedback between chironomids and algae creates net mutualism between benthic primary consumers and producers. *Ecology* **2017**, *98*, 447–455. [CrossRef]

59. Hamm, C.E.; Merkel, R.; Springer, O.; Jurkojc, P.; Maier, C.; Prechtel, K.; Smetacek, V. Architecture and material properties of diatom shells provide effective mechanical protection. *Nature* **2003**, *421*, 841–843. [CrossRef]

60. Fenoglio, S.; Bo, T.; Tierno de Figueroa, J.M.; Cucco, M. Nymphal growth, life cycle, and feeding habits of *Potamanthus luteus* (Linnaeus, 1767) (Insecta: Ephemeroptera) in the Bormida River, Northwestern Italy. *Zool. Stud.* **2008**, *47*, 185–190.

61. Botana, L.M. *Seafood and Freshwater Toxins. Pharmacology, Physiology and Detection*; CRC Press: New York, NY, USA, 2000; p. 1215.

62. Dhar, B.C.; Cimarelli, L.; Singh, K.S.; Brandi, L.; Brandi, A.; Puccinelli, C.; Marcheggiani, S.; Spurio, R. Molecular detection of a potentially toxic diatom species. *Int. J. Environ. Res. Public Health* **2015**, *12*, 4921–4941. [CrossRef]

63. Violi, J.P.; Facey, J.A.; Mitrovic, S.M.; Colville, A.; Rodgers, K.J. Production of β-methylamino-L-alanine (BMAA) and its isomers by freshwater diatoms. *Toxins* **2019**, *11*, 512. [CrossRef]

64. Falasco, E.; Bona, F. Recent findings regarding non-native or poorly known diatom taxa in north-western Italian rivers. *J. Limn.* **2013**, *201*, 35–51. [CrossRef]

65. Ladrera, R.; Goma, J.; Prat, N. Effects of *Didymosphenia geminata* massive growth on stream communities: Smaller organisms and simplified food web structure. *PLoS ONE* **2018**, *13*, e0193545. [CrossRef]

66. Bray, J.P.; Kilroy, C.; Gerbeaux, P.; Burdon, F.J.; Harding, J.S. Ecological processes mediate the effects of the invasive bloom-forming diatom *Didymosphenia geminata* on stream algal and invertebrate assemblages. *Hydrobiologia* **2020**, *847*, 177–190. [CrossRef]

67. Wahl, M. Epibiosis. In *Marine Hard Bottom Communities*; Wahl, L., Ed.; Springer: Boston, MA, USA, 2009; pp. 61–72.

68. Falasco, E.; Bo, T.; Ghia, D.; Gruppuso, L.; Bona, F.; Fenoglio, S. Diatoms prefer strangers: Non-indigenous crayfish host completely different epizoic algal diatom communities from sympatric native species. *Biol. Inv.* **2018**, *20*, 2767–2776. [CrossRef]

69. Wahl, M.; Goecke, F.; Labes, A.; Dobretsov, S.; Weinberger, F. The second skin: Ecological role of epibiotic biofilms on marine organisms. *Front. Microbiol.* **2012**, *3*, 292. [CrossRef]

70. Robinson, N.J.; Majewska, R.; Lazo-Wasem, E.A.; Nel, R.; Paladino, F.V.; Rojas, L.; Zardus, J.D.; Pinou, T. Epibiotic diatoms are universally present on all sea turtle species. *PLoS ONE* **2016**, *11*, e0157011. [CrossRef]

71. Wetzel, C.E.; Van de Vijver, B.; Cox, E.J.; Bicudo, D.D.C.; Ector, L. *Tursiocola podocnemicola* sp. nov., a new epizoic freshwater diatom species from the Rio Negro in the Brazilian Amazon Basin. *Diatom Res.* **2012**, *27*, 1–8. [CrossRef]

72. Abbott, L.L.; Bergey, E.A. Why are there few algae on snail shells? The effects of grazing, nutrients and shell chemistry on the algae on shells of *Helisoma trivolvis*. *Freshwat. Biol.* **2007**, *52*, 2112–2120. [CrossRef]

73. Wujek, D.E. The first occurrence of the chrysophyte alga *Amphirhiza epizootica* from North America. *Mich. Bot.* **2006**, *45*, 197–200.

74. Winemiller, K.; Winsborough, B. Occurrence of epizoic communities on the parasitic copepod *Lernaea carassii* (Lernaeidae). *Southwest. Nat.* **1990**, *35*, 206–210. [CrossRef]

75. Fuelling, L.J.; Adams, J.A.; Badik, K.J.; Bixby, R.J.; Caprette, C.L.; Caprette, H.E.; Hall, M.M. An unusual occurrence of *Thorea hispida* (Thore) *Desvaux chantransia* on rusty crayfish in West Central Ohio. *Nova Hedwig.* **2012**, *94*, 355–366. [CrossRef]

76. Bergey, E.A.; Resh, V.H. Interactions between a stream caddisfly and the algae on its case: Factors affecting algal quantity. *Freshwat. Biol.* **1994**, *31*, 153–163. [CrossRef]

77. Wujek, D.E. Epizooic diatoms on the cerci of Ephemeroptera (Caenidae) naiads. *Great Lakes Entomol.* **2013**, *46*, 117–120.

78. Stewart, K.W.; Schlichting, H.E., Jr. Dispersal of algae and protozoa by selected aquatic insects. *J. Ecol.* **1966**, *54*, 551–562. [CrossRef]

79. Revill, D.L.; Stewart, K.W.; Schlichting, H.E. Passive dispersal of viable algae and protozoa by certain craneflies and midges. *Ecology* **1967**, *48*, 1023–1027. [CrossRef]

80. Göthe, E.; Angeler, D.G.; Gottschalk, S.; Löfgren, S.; Sandin, L. The influence of environmental, biotic and spatial factors on diatom metacommunity structure in Swedish headwater streams. *PLoS ONE* **2013**, *8*, e72237.

81. Burgazzi, G.; Bolpagni, R.; Laini, A.; Racchetti, E.; Viaroli, P. Algal biomass and macroinvertebrate dynamics in intermittent braided rivers: New perspectives from instream pools. *River Res. Appl.* **2020**. [CrossRef]

82. Wisz, M.S.; Pottier, J.; Kissling, W.D.; Pellissier, L.; Lenoir, J.; Damgaard, C.F.; Dormann, C.F.; Forchhammer, M.C.; Grytnes, J.-A.; Guisan, A.; et al. The role of biotic interactions in shaping distributions and realised assemblages of species: Implications for species distribution modelling. *Biol. Rev.* **2013**, *88*, 15–30. [CrossRef]

 *water*

*Article*

# Key Determinants of Freshwater Gastropod Diversity and Distribution: The Implications for Conservation and Management

**Mi-Jung Bae [1] and Young-Seuk Park [2],*** 

[1] Biodiversity Research Team, Freshwater Biodiversity Research Bureau, Nakdonggang National Institute of Biological Resources, Sangju, Gyeongsangbuk-do 37242, Korea; mjbae@nnibr.re.kr

[2] Department of Biology, Kyung Hee University, Dongdaemun, Seoul 02447, Korea

* Correspondence: parkys@khu.ac.kr; Tel.: +82-2-961-0946

Received: 2 June 2020; Accepted: 30 June 2020; Published: 4 July 2020

**Abstract:** Freshwater organisms are facing threats from various natural and anthropogenic disturbances. Using data sampled on a nationwide scale from streams in South Korea, we identified the crucial environmental factors influencing the distribution and abundance of freshwater gastropods. We used nonmetric multidimensional scaling and the random forest model to evaluate the relationships between environmental factors and gastropod assemblages. Among the 30 recorded species, two invasive gastropod species (*Pomacea canaliculata* and *Physa acuta*) have enlarged their distribution (10.4% and 57.3% frequency of occurrence, respectively), and were found to be widespread in streams and rivers. Our results revealed that the most influential factor in the distribution of gastropod assemblages was the ratio of cobble (%) in the substrate composition, although meteorological and physiographical factors were also important. However, the main environmental factors influencing species distribution varied among species according to habitat preference and environmental tolerance. Additionally, anthropogenic disturbance caused a decrease in the distribution of endemic species and an increase in the spatial distribution of invasive species. Finally, the results of the present study provide baseline information for planning successful strategies to maintain and conserve gastropod diversity when facing anthropogenic disturbance, as well as understanding the factors associated with the establishment of invasive species.

**Keywords:** snails; endemic species; invasive species; random forest model; multivariate analysis; partial dependence analysis

## 1. Introduction

Freshwater gastropods comprise 5% of the global gastropod fauna but are facing a disproportionately high degree of threat according to the 2019 IUCN Red List of Threatened Species (http://www.redlist.org, Cambridge, UK) [1]. Furthermore, 74% of gastropod species have been classified as vulnerable, endangered, threatened, or already extinct in the United States and Canada [2]. Similarly, in mollusks, 40% of freshwater bivalve species in the world are near threatened, threatened, or extinct [3]. In South Korea, two rare species (*Clithon retropictus* and *Koreanomelania nodifila*) are listed in the Red Book as endangered, and five species (*K. nodifila, Koreoleptoxis globus ovalis, Semisulcospira coreana, Semisulcospira forticosta*, and *Semisulcospira tegulata*) are endemic in the database of the National Aquatic Ecological Monitoring Program (National Institute of Environmental Research, Incheon, South Korea) [4]. These threats are of particular global concern because most freshwater gastropods are endemic with small geographic ranges [2,5–7].

The distribution and structure of gastropod assemblages are influenced by various environmental factors [8,9]. Across a large-scale area that includes several basins, climate-related factors, such as temperature and precipitation as well as physiographical factors, are important in influencing the structure of freshwater gastropod assemblages [10]. Other environmental factors, such as hydrological, physicochemical, and biological factors, are important within the same climate region. Differences in vegetation, land use, and flood disturbance are important in explaining variation in species composition among streams [11,12]. In addition, within-stream and/or -microhabitat differences (e.g., differences between flow regimes, substrate composition, and riparian vegetation) contribute to variations among gastropod assemblages [13–15].

Gastropods are sessile and have a very limited ability to avoid unfavorable environments, making it difficult to recover the heterogeneity of a freshwater ecosystem once it has been disrupted [10,16]. Therefore, research on the multiscale regulation of gastropod assemblage structures in freshwater ecosystems is the first step in planning a successful strategy for either conserving or restoring freshwater diversity. However, most studies on species distribution and assemblage structure have had a small-scale focus or have been mainly conducted in lentic habitats [17,18], or considered only a limited number of environmental factors [19–21].

Therefore, the present study aimed to identify environmental factors influencing gastropod distribution patterns, focusing on lotic habitats on a nationwide scale (South Korea). Specifically, we tested two hypotheses. First, large-scale factors, including temperature-related and physiographical factors, are more influential than other factors in our research area to determine gastropod assemblage structure. Second, major influential environmental factors differ depending on species. Finally, we considered strategies for the conservation and management of freshwater gastropods based on our results.

## 2. Materials and Methods

### 2.1. Ecological Data

Gastropod data were obtained from the database of the NAEMP (http://water.nier.go.kr/). Since 2008, NAEMP has conducted nationwide surveys of the freshwater organisms of South Korea, including periphyton, benthic macroinvertebrates, and fish, together with associated environmental factors, twice per year (i.e., spring and autumn, when natural disturbances such as heavy rain and drought are at a minimum). The river systems in South Korea form five major river basins (i.e., the Han, Nakdong, Yeongsan, Seomjin, and Geum River basins). Among these, the Han River basin (basin area: 41,957 km$^2$) in the north of South Korea is the largest, occupying one-third of the country. The Nakdong River basin (31,785 km$^2$) is in the southeast, and the Geum River basin (17,537 km$^2$) is in the midwest. Lastly, the Yeongsan River (3,467 km$^2$) and Seomjin River (4,912 km$^2$) basins are located close to each other in the southwest of South Korea [4,22–24].

Gastropods were collected at 714 sampling sites from all the South Korean river basins twice per year from 2008 to 2013, using a Surber net (30 × 30 cm, 1 mm mesh size, Table 1). Three replicates were collected from each sampling site at each sampling time and then were transformed into abundance/m$^2$ for the further analyses, based on NAEMP guidelines [25]. Detailed information regarding the sampling protocol is given in previous studies [26–28]. The samples were preserved in 95% ethanol in the field, and then placed in 70% ethanol in the laboratory. They were sorted and identified, mostly to species level, and the number of individuals per species was counted using naked-eye or microscope examination [4].

**Table 1.** Average values and standard deviations (SD) of the environmental variables characterizing the sampling sites.

| Environmental variable | Abbreviation | N | Average | S.D. |
|---|---|---|---|---|
| *Meteorology* | | | | |
| Annual average temperature (°C) | Ave_temp | 714 | 11.2 | 1.3 |
| Average temperature in August (°C) | Aug_temp | 714 | 25.0 | 1.2 |
| Average temperature in January (°C) | Jan_temp | 714 | −4.0 | 1.9 |
| Thermal range (°C) | Thermal_range | 714 | 29.0 | 1.5 |
| Annual precipitation (mm) | | 714 | 1112 | 142 |
| *Physiography* | | | | |
| Altitude (m) | | 714 | 114 | 130 |
| Slope (°) | | 714 | 4.2 | 5.9 |
| Distance from source (km) | DFS | 714 | 50.8 | 80.1 |
| Stream order | Str_order | 714 | 4 | 1 |
| *Land use* | | | | |
| Urban (%) | | 714 | 18.7 | 24.9 |
| Agriculture (%) | | 714 | 44.0 | 30.3 |
| Forest (%) | | 714 | 30.0 | 30.7 |
| *Hydrology* | | | | |
| Water width (m) | W_width | 711 | 64.9 | 98.1 |
| Water depth (cm) | Ave_depth | 714 | 33.0 | 18.8 |
| Water velocity (cm/s) | Ave_velocity | 714 | 38.8 | 23.8 |
| Riffle (%) | | 714 | 18.1 | 17.3 |
| Run (%) | | 714 | 70.9 | 20.8 |
| Pool (%) | | 714 | 11.0 | 15.7 |
| *Substrate* | | | | |
| Silt (%) | | 714 | 3.9 | 9.3 |
| Clay (%) | | 714 | 8.5 | 12.1 |
| Sand (%) | | 714 | 24.7 | 18.4 |
| Small pebble (%) | S_pebble | 714 | 18.4 | 8.2 |
| Pebble (%) | | 714 | 20.7 | 11.0 |
| Cobble (%) | | 714 | 18.1 | 12.4 |
| Boulder (%) | | 714 | 5.5 | 7.7 |
| *Water quality* | | | | |
| Biochemical oxygen demand (mg/L) | BOD | 713 | 1.39 | 0.48 |
| Total nitrogen (mg/L) | TN | 713 | 2.51 | 0.78 |
| Ammonia nitrogen (mg/L) | NH3N | 707 | 0.05 | 0.03 |
| Nitrate nitrogen (mg/L) | NO3N | 714 | 1.47 | 0.42 |
| Total Phosphorus (mg/L) | TP | 694 | 0.05 | 0.02 |
| Orthophosphate (mg/L) | PO4P | 679 | 0.02 | 0.01 |
| Chlorophyll *a* (µg/L) | Chl.a | 714 | 1.53 | 0.67 |
| Dissolved oxygen (mg/L) | DO | 714 | 9.25 | 1.17 |
| pH | | 714 | 7.78 | 0.34 |
| Electric conductivity (µS/cm) | Conductivity | 709 | 176.3 | 78.5 |

### 2.2. Data Analyses

We analyzed the data in three steps. First, we classified the sampling sites by conducting a hierarchical cluster analysis (CA) based on gastropod abundance. The CA was calculated based on the Bray–Curtis dissimilarity with Ward's linkage method [29], using the 'vegan' package [30] in R [31]. Then, multiresponse permutation procedures (MRPP) was considered to evaluate the significant differences among the clusters defined through CA. We defined the indicator species in each cluster using an indicator species analysis [32]. The indicator species was selected based on the indicator value (IndVal), by considering relative species abundance and its associated relative frequency of occurrence within the defined clusters. The IndVal range was from 0 to 100 (all individuals of a species are included only within a single cluster). Species with a statistically significant ($p < 0.05$) IndVal higher than 25% were selected as indicator species [32]. A site randomization procedure that reallocates samples among sample groups (9999 permutations) was used to test for significance. Indicator species analysis was performed using the 'indval' function in the 'labdsv' package [33] in R [31].

Second, to describe the gastropod assemblage patterns, we applied nonmetric multidimensional scaling (NMDS) based on the Bray–Curtis dissimilarity between sampling sites, using the 'vegan' package [34] in R [31]. In order to identify the best NMDS solution (i.e., the lowest STRESS value), the 'metaMDS' function was applied. Then, we used the 'envfit' function to evaluate the relationships between gastropod assemblages and environmental factors [34,35]. All the analyses related to NMDS were conducted using the 'vegan' package in R [31].

Lastly, a random forest (RF) model was used to predict the distribution of gastropod species based on gastropod abundance and environmental variables, and to evaluate the contribution of each environmental variable to species distribution. The RF model, a machine learning model, is computed using a combination of a large set of decision trees [4,36], and does not require assumptions, such as linear or nonlinear relationships between predictors (environmental variables in the present study) and response factors (gastropod species) [27,37,38]. We used the 'randomForest' package [39] in R [31], with the three default training parameters: ntree (number of trees = 500), mtry (number of variables = 3), and node size (5). The importance of environmental factors to gastropod distribution was computed based on the mean decrease in accuracy, and importance values were then rescaled from 0 to 100 [40,41]. The RF model was applied to dominant gastropod species (here, more than 10% occurrence frequency, Table 2). Then, we used partial dependence analysis [42] to evaluate the relationship between environmental factors and 11 gastropod species presenting more than a 10% frequency of occurrence in all sampling sites.

**Table 2.** Gastropod species and their frequency of occurrence (%) in the dataset. Species with a frequency of occurrence of over 10% are indicated in bold.

| Order | Family | Species | Abbreviation | Frequency of Occurrence (%) [d] |
|---|---|---|---|---|
| Lepetellida | Trochidae | *Monodonta neritoides* | Mo_ne | 0.14 |
| Mesogastropoda | Viviparidae | ***Cipangopaludina chinensis malleata*** | Ci_ch | **14.15** |
| | | *Cipangopaludina japonica* | Ci_ja | 1.26 |
| | | *Sinotaia quadrata* | Si_qu | 0.14 |
| | Ampullariidae | ***Pomacea canaliculata*** [c] | Po_ca | **10.36** |
| | Bithyniidae | *Gabbia misella* | Ga_mi | 5.60 |
| | | *Parafossarulus manchouricus* | Pa_ma | 3.78 |
| | Assimineidae | *Assiminea japonica* | As_ja | 0.84 |
| | | *Assiminea lutea* | As_lu | 0.14 |
| | Stenothyridae | *Stenothyra glabra* | St_gl | 8.12 |
| | Netritidae | *Clithon retropictus* [a] | Cl_re | 1.54 |
| | Pleuroceridae | *Koreanomelania nodifila* [a,b] | Ko_no | 4.90 |
| | | *Koreanomelania paucicincta* | Ko_pa | 0.70 |
| | | *Koreoleptoxis globus* [b] | Ko_gl | 2.38 |
| | | *Koreoleptoxis globus ovalis* | Ko_gl_o | 1.26 |
| | | ***Semisulcospira coreana*** | Se_co | **19.05** |
| | | ***Semisulcospira forticosta*** | Se_fo | **26.75** |
| | | ***Semisulcospira gottschei*** | Se_go | **23.53** |
| | | ***Semisulcospira libertine*** | Se_li | **49.02** |
| | | ***Semisulcospira tegulata*** | Se_te | **14.71** |
| | | *Semisulcospira paucicincta* | Se_pa | 0.70 |
| Basommatophora | Lymnaeidae | *Austropeplea ollula* | Au_ol | 4.76 |
| | | *Fossaria truncatula* | Fo_tr | 0.56 |
| | | ***Radix auricularia*** | Ra_au | **44.40** |
| | Physidae | ***Physa acuta*** [c] | Ph_ac | **57.28** |
| | Planorbidae | ***Gyraulus convexiusculus*** | Gy_co | **25.07** |
| | | ***Hippeutis cantori*** | Hi_ca | **21.57** |
| | | *Polypylis hemisphaerula* | Po_he | 3.36 |
| | Ancylidae | *Laevapex nipponicus* | La_ni | 1.96 |
| | Succineidae | *Oxyloma hirasei* | Ox_hi | 2.66 |

[a]: endangered species, [b]: endemic species, [c]: invasive species, and [d]: percentage of observed sampling sites.

The abundance of each gastropod species was averaged at each site after pooling yearly data and transformed with natural logarithm to reduce variation in abundance prior to further analyses. In environmental factors, the extremes and outliers were removed before analyzing the data and the values were averaged after pooling yearly data like gastropod abundance.

## 3. Results

Thirty species, belonging to three orders and 13 families, were recorded in the study area (Table 2). *Physa acuta*, an invasive species, was the most commonly observed (57.3% occurrence frequency of all the sites), followed by *Semisulcospira libertina* (49.0%), *Radix auricularia* (44.4%), and *S. forticosta* (26.8%, Figure 1 and Table 2). Two species, *Clithon retropictus* and *Koreanomelania nodifila*, listed in the Red Book of Korea (National Institute of Biological Resources, Incheon, South Korea) as endangered had low occurrence frequencies (1.5% and 4.9%, respectively), and *Pomacea canaliculata*, an invasive species, was recorded in 10.4% of the sites.

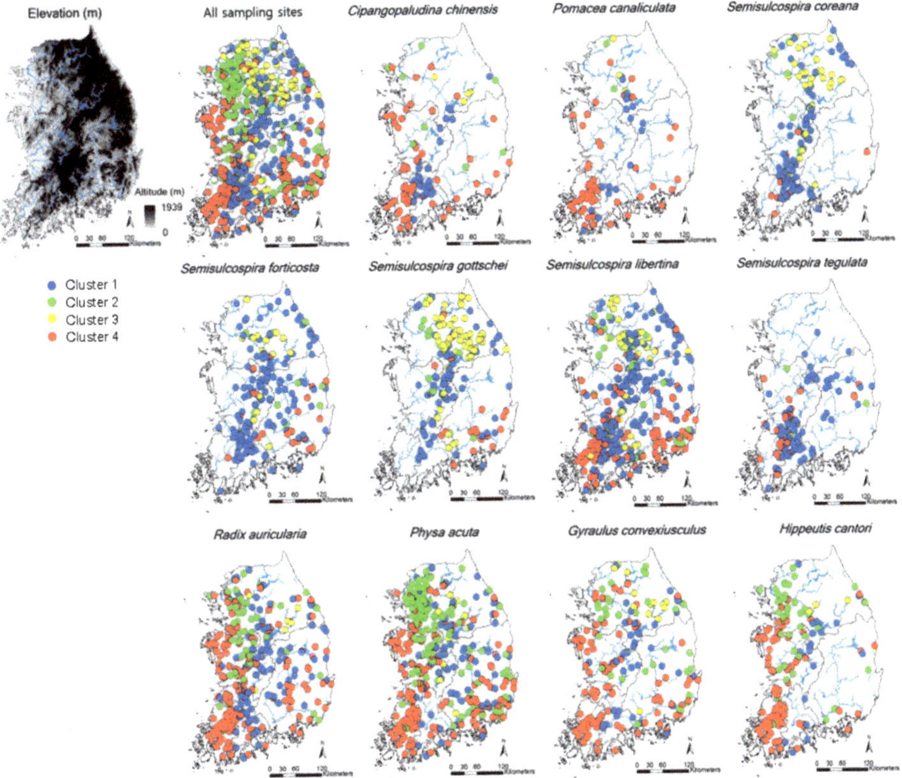


**Figure 1.** Sampling sites and occurrence patterns of 11 gastropod species in Korea that present more than a 10% frequency of occurrence. Classification of sampling sites were defined in Figure 2. The clusters reflected the differences of environmental condition at each sampling site.

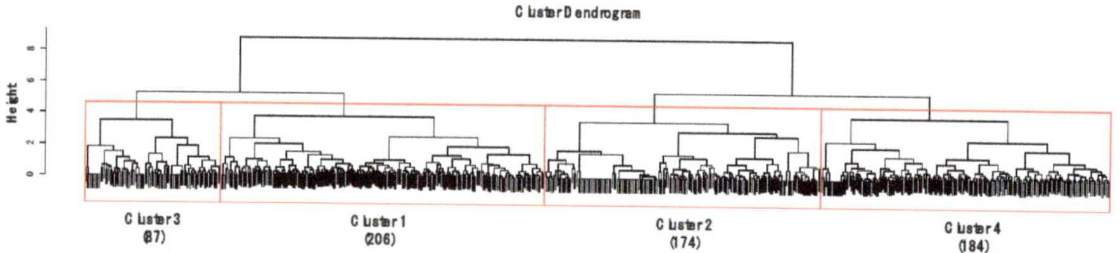

**Figure 2.** Dendrogram of cluster analysis based on gastropod assemblages using Ward's linkage method with the Bray–Curtis dissimilarity. Height on the y-axis indicates distances of merging clusters, reflecting the distance between the samples. Numbers in parenthesis represent the number of sampling sites in each cluster.

CA classified the sites into four clusters (1–4) based on similarities in gastropod assemblage composition (Figure 2) and four clusters were significantly different based on MRPP ($A = 0.18$–$0.21$, $p < 0.05$). A total of 21 species were selected as indicator species for four clusters based on their IndVal ($p < 0.05$, Table S1). Cluster 4 contained the highest number of indicator species (10), followed by Cluster 1 (8), Cluster 3 (2), and Cluster 2 (1). In Clusters 2 and 4, the species (i.e., *Hippeutis cantori* and *Austropeplea ollula*) mainly found in organic enriched streams and/or invasive species (i.e., *P. canaliculata* and *P. acuta*) were selected as an indicator species. On the other hand, in Clusters 1 and 3, the species (i.e., *S. gottschei*) mostly found in the less disturbed area and endangered and/or endemic species (i.e., *K. globus* and *K. nodifila*) were mainly selected.

NMDS also showed differences in gastropod assemblage composition (Figure 3), reflecting the classification of sampling sites in CA. After the NMDS ordination with gastropod assemblage, environmental variables were visualized with sampling sites and gastropod taxa with biplot. We selected the first three NMDS axes based on the Kruskal's stress value (the first three axes stress = 13.8). Clusters 1 and 3 and Clusters 2 and 4 were differentiated on the NMDS 1, whereas Clusters 2 and 3 and Clusters 1 and 4 were divided on the NMDS 2 (Figure 3a). However, the differentiation of clusters was not clear on NMDS 3. NMDS 1 reflected the gradient of water quality, whereas NMDS2 presented the gradient of temperature. On the ordination with NMDS 1 and NMDS 2, sampling sites with high values of cobble (%) and altitude were located on the left side of NMDS 1, whereas sites with high values of TN, conductivity, BOD, and TP were located on the right side of NMDS 1 (Figure 3a). Sampling sites with high values of average temperature in January were located on the lower part of NMDS 2, whereas sampling sites with high values of TP and conductivity were located on the lower part of NMDS 3. Species which prefer riffle areas with a large-sized substrate and good water quality, such as *Koreoleptoxis globus ovalis*, *K. nodifila*, *S. coreana*, and *S. forticosta*, were on the left section of NMDS 1 (Figure 3b). Conversely, the high values of TN, conductivity, BOD, and TP strongly influenced the distributions of *R. auricularia*, *P. acuta*, *Gyraulus convexiusculus*, and *H. cantori*. The average temperature in January was influential to the distributions of *P. canaliculata* and *C. retropictus*.

Distributions of species were well predicted by the RF models, with a high prediction power ranging from 0.97 to 0.99 (Figure 4). Overall, meteorological and physiographical variables were included in the main factors influencing the distribution of gastropod species. However, the contribution of other environmental factors for predicting species distribution differed depending on species. For example, conductivity was the most important factor (100) for predicting the occurrence of *P. acuta*, followed by cobble (94), and water depth (65). TN (100), silt (89), and riffle (86) were important for predicting *S. libertine* abundance. Average temperature in January (100) was the most important factor for *P. canaliculata*, an invasive species in South Korea, followed by water velocity (90), annual average temperature (80), and distance from the source (78). The ratio (%) of cobble (100) and TN (77) were influential in determining the occurrence of *S. coreana*.

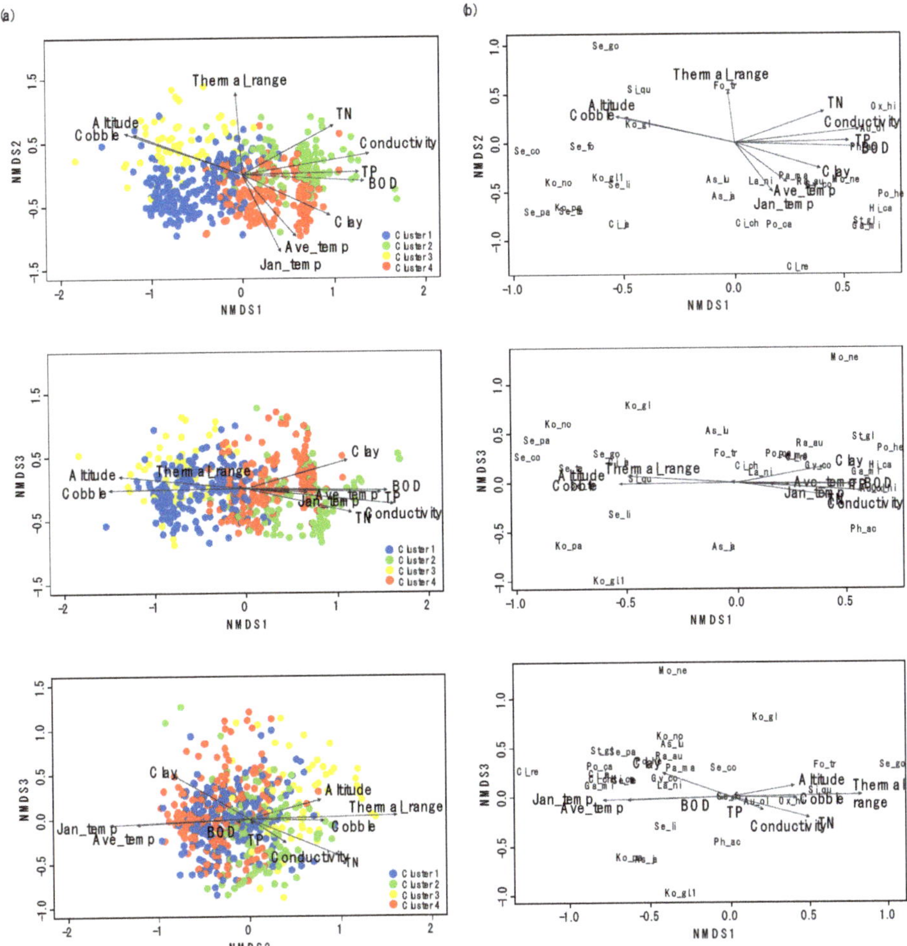

**Figure 3.** Nonmetric multidimensional scaling (NMDS) ordination of sampling sites (**a**) with gastropod assemblages (**b**). The first three axes (Kruskal's stress = 13.8) were used to visualize the ordination. Arrows correspond to environmental variables significantly related to assemblage composition (only environmental variables with $R^2 > 0.2$ are presented). Arrow length is proportional to the correlation magnitude ($r$). The abbreviations for the environmental variables and species names are explained in Tables 1 and 2, respectively.

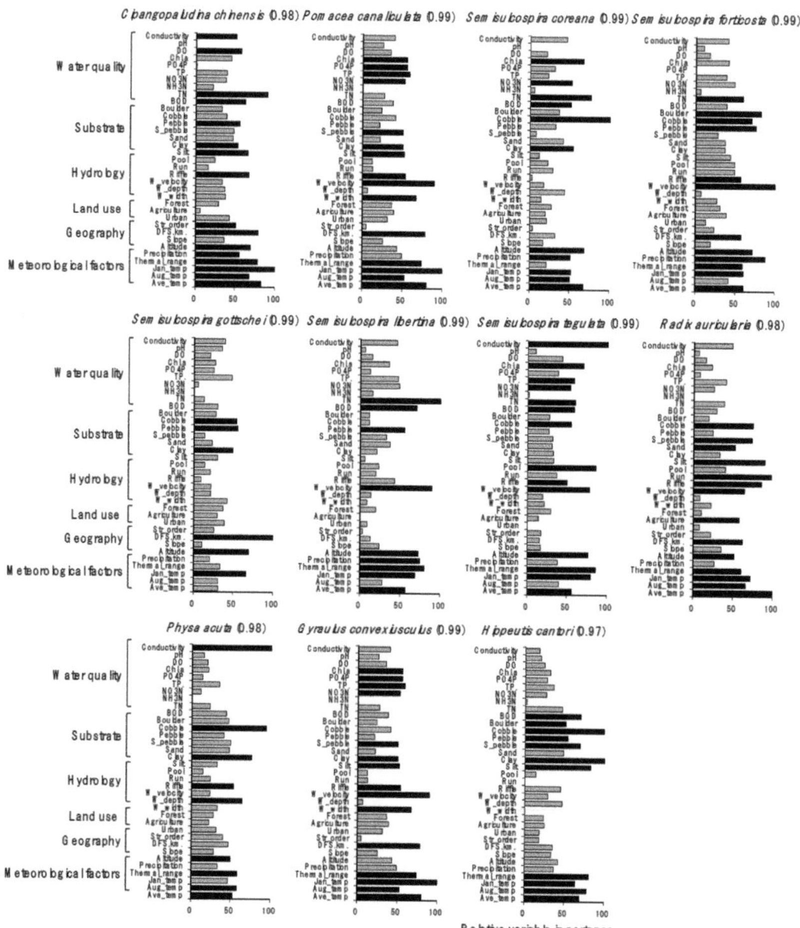

**Figure 4.** Relative importance of environmental variables for predicting the distribution of gastropod species in the random forest model. Numbers in parentheses indicate the predictability of the model for each species. black bar: ≥50 of the relative importance value and gray bar: <50 of the relative importance value.

The partial dependence plot displayed that each gastropod species responded differently to the environmental factors in the RF models (Figure 5 and Figure S1). We visualized two variables showing a strong gradient in the NMDS plot: altitude on NMDS 1 and the average temperature in January on NMDS 2 (Figure 3). Most of genus *Semisulcospira*, *C. chinensis*, and *P. canaliculata* displayed increase of their abundance as increase of altitude, whereas *P. acuta* was abundant mostly at low altitude (≤160 m, Figure 5a). The high abundance of *P. acuta* was also related to high concentrations of TN (≥1.8 mg/L, Figure S1). Meanwhile, responding to temperature, the abundance of *P. canaliculata* increased with an increasing average temperature in January above −2.0 °C, whereas abundances of *S. gottschei* and *S. libertina* were high at a lower temperature in January (−4.2 °C and −2.2 °C, respectively, Figure 5b).

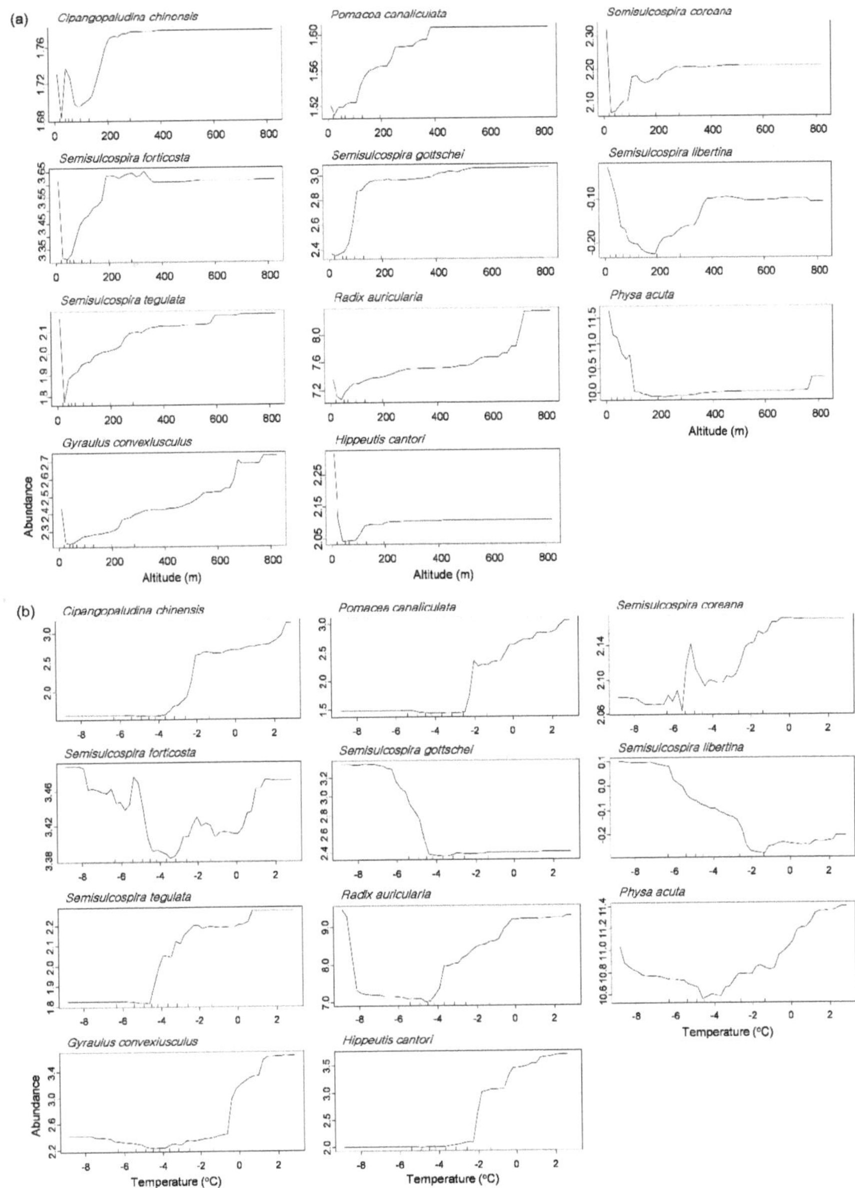

**Figure 5.** Partial dependence plots of 11 abundant species responding to altitude (**a**) and average temperature in January (**b**) in the random forest model.

## 4. Discussion

### 4.1. Environmental Factors Influencing Gastropod Assemblages

The structure and distribution of the gastropod assemblage were differentiated by various environmental factors in our study. Meteorological (e.g., temperature-related factors), physiographical (e.g., altitude), substrate composition (e.g., the ratio of cobble) and water quality (e.g., conductivity, TN,

TP, and BOD) gradients were important in structuring gastropod assemblages. Among them, the ratio of cobble (%) was the most influential factor in gastropod assemblage. Substrate composition is closely related to habitat complexity and resource availability (e.g., the number of algae, aquatic macrophytes, and microorganisms), especially for gastropods [43]. For instance, the distribution of *Semisulcospira* which mainly feed on periphytic algae attached to large-sized substrate materials, such as boulders and cobbles, was highly related to the ratio of cobble (%) [44]. In addition to substrate composition, gastropod assemblages showed differentiation from the least polluted to anthropogenically disturbed streams (i.e., the gradient of water quality). Species such as *R. auricularia*, *P. acuta*, *G. convexiusculus*, and *H. cantori* occurred frequently at sites with high values of conductivity, BOD, TN, and TP. Because these species can assimilate atmospheric air through their vascularized mantle cavity, they can thrive under harsh conditions [45], for example, in areas with silt substrates and high conductivity [46]. Physidae, in particular, have a high tolerance to organic pollution [47]; the occurrence of the genus *Physa* is associated with relatively higher values of pH, calcium hardness, total hardness, total alkalinity, conductivity, and total dissolved solids than other species [48,49].

### 4.2. Conservation of Gastropoda Species

In South Korea, the number of species in Pleuroceridae (10 species) was the highest among gastropod taxa, especially in the genus *Semisulcospira* (6 species). The distribution of *Semisulcospira* was related to cobble percentage in substrate composition. This might be linked to food resources and habitat characteristics, as *Semisulcospira* mainly is a scraper feeding on periphytic algae attached to large-sized substrates like boulders and cobbles without dense canopy cover in riparian areas (Karube et al., 2012). *Semisulcospira* is a key food resource for many freshwater organisms, especially for species with high conservation priority in South Korea, such as Spotted barbel (*Hemibarbus mylodon*; fish, Korean National Monument No. 259 and endemic species) as well as firefly (*Luciola lateralis* and *Lychnuris rufa*, insect). Especially, the main population of the latter two species (i.e., *L. lateralis* and *L. rufa*) inhabit only within a confined area in Muju-gun (Jeollabuk-do, South, Korea) with their food sources (i.e., *Semisulcospira*). Therefore, this area is designated as the Korean National Monument (No 332) to conserve their population as well as *Semisulcospira* population. However, in spite of their important roles including the critical food sources in the freshwater ecosystem, *Semisulcospira* has been excessively collected by human beings for domestic consumption (i.e., snail soup) because *Semisulcospira* is well known to people about its the high-protein food sources. In addition, collecting *Semisulcospira* in streams is one of the general and popular family leisure activities in Korea, especially in the summer season. However, because people cannot distinguish the genus *Semisulcospira* into species level and just collect them without considering their ecological importance, resulting in the dramatic reduction of its abundance and diversity.

Currently, the habitable area for most species, including endangered species, is being consistently lost [50]. Endangered and/or endemic species distributed in only one basin should, therefore, be selected as the first priority for management and conservation [51]. Fortunately, in our study, no endemic and/or endangered species were found to be inhabiting a single watershed, with limited distribution. However, anthropogenic disturbances could threaten gastropod species, especially those that are endemic and/or endangered, or which have a limited distribution. For instance, we found that the occurrence frequency of endemic and/or endangered species was less than 5%. Among three species recorded as endemic and/or endangered, *C. nodifila* attracts a particularly high conservation concern. This is because this species currently has a limited and narrow distribution, being found only in some of the southern parts of South Korea (1.5% frequency of occurrence), such as the Yeongsan and Seumjin River basins, which are in the least disturbed area with a shallow water depth and low current velocity. *C. retropictus* is distributed mainly within tropical and subtropical and some temperate regions [52]; the distribution of these taxa is, therefore, also sensitive to low temperature and large thermal ranges. Moreover, bank and dam construction could destroy the habitable environment of *C. retropictus* in various ways. This species is designated within the second grade of endangered wild fauna and flora species by the Ministry of Environment in the Republic of Korea, and its populations require active conservation and management.

*K. nodifila* is another endangered species that is endemic to South Korea. In our study, it was found mainly to be restricted (4.9% frequency of occurrence) to the least disturbed upstream areas in the northern part of South Korea (i.e., in the Imjin, Hantan, and Dong Rivers). The distribution of *K. nodifila* is strongly influenced by small variations in stream habitat conditions. For instance, this species prefers natural habitats that have a high water velocity, riffle (%), and the ratio of cobble (%), as well as good water quality. However, its original habitat has suffered continuous disruption and habitat loss, due to the construction and reorganization of flood control dams and weirs. These constructions alter the flow regime from lotic to lentic, a factor that is influential in the occurrence of this species. Furthermore, *K. nodifila* may be gathered with *Semisulcospira* species when the latter is collected during the snail hand-picking season for use in cooking snail soup and/or leisure activities in Korea. This is because it is difficult for the general public to distinguish *Semisulcospira* species from *K. nodifila*. In this sense, ongoing public relations and education need to cover the importance and key characteristics of *K. nodifila* to protect the species from excessive collection.

*4.3. Management of Invasive Species*

In this study, the distribution of *P. canaliculata* (10.4% frequency of occurrence) was determined by two critical factors: average temperature in January and percentage of agricultural area (Figure 4). *P. canaliculata* is native to tropical areas of South America and is one of the world's 100 worst invasive species [53]. In South Korea, this species has been introduced as a food resource, and as a herbicide substitute to control weeds in environmentally friendly paddy fields. In fact, the increased usage of this snail is mostly due to its high efficiency in controlling weeds (99%) [54]. *P. canaliculata* was introduced to South Korea on the assumption that it would be unable to overwinter there because of the low winter temperatures [55]. Currently, however, 30 years after its first introduction to South Korea, there are frequent reports of *P. canaliculata* overwintering in open freshwater ecosystems, especially in the southern part of South Korea [56]. In addition, the geographic range over which *P. canaliculata* can overwinter is continuously increasing due to global warming and the biological adaptation of this invasive species, resulting in severe impacts on aquatic ecosystems [57]; based on our database, it is already found in all the South Korean river basins. The invasion of Thailand's natural wetlands by *P. canaliculata* is causing aquatic plants to disappear from riparian systems, resulting in high nutrient concentrations and phytoplankton biomass [58]. Therefore, there has been a complete shift in the state of the ecosystem and functions in areas where *P. canaliculata* has become established. Ongoing and systematic management of this species is, therefore, essential if it is to be eradicated and to prevent its further expansion and establishment in the aquatic ecosystems of South Korea.

In addition, there are many studies reporting that land use disturbance alters landscapes, ecosystem structure, and functions [40]. The continuous increase in agricultural areas and urbanization is leading to an increase in the introduction of invasive species and homogenization of gastropod assemblages [59]. In our study, species indicating greater tolerance of organic pollution and urbanization possessed the highest frequencies of occurrence. In particular, *P. acuta* had a frequency of occurrence of 57.3%, the highest of all species in our study; similarly, *P. canaliculata*, which has a high tolerance to organic enrichment, had a frequency of occurrence of 10.4%. Therefore, habitat degradation can induce changes in the structure and distribution of gastropod assemblages and cause the distribution of both tolerant and invasive species to expand.

## 5. Conclusions

Our evaluation of the two questions examined in this study revealed the following results. (1) The ratio of cobble in the substrate composition was the most influential factor in gastropod assemblage distribution on a national scale. (2) Nonetheless, the major environmental factors influencing the distribution of each species varied according to habitat preference and environmental tolerance. This study is the first to quantify the distribution ranges of all gastropod species on the national scale, and to evaluate the influential factors determining that distribution based on a modeling approach. Even though no endemic and/or endangered species were found inhabiting only a single basin, they nonetheless have a limited distribution (less than 5%). Furthermore, invasive gastropod species have extended their distribution (*P. canaliculata*, 10.4%; and *P. acuta*, 57.3%) and are easily found within all the river basins. The habitats occupied by gastropods have been continuously disrupted by various factors, resulting in reductions to available suitable habitat. Local pollution, hydrologic alteration, agriculture, global warming, and the introduction of invasive species have had severe impacts on aquatic ecosystems. Detailed information on gastropod assemblages and the factors influencing their assemblage structure and distribution is, therefore, required for the successful conservation of aquatic gastropods. In particular, the distribution and abundance of invasive and endangered species should be evaluated to prevent ecosystem disruption and enhance species conservation strategies. Finally, the results of the present study would contribute to the development of adequate and systematic management policies for the conservation and management of freshwater gastropods.

**Supplementary Materials:** The following are available online at http://www.mdpi.com/2073-4441/12/7/1908/s1, Table S1: Indicator species selected in four clusters based on the Indval. Only taxa with significant values are shown. Figure S1: Partial dependence plots of 11 abundant species responding to (a) total nitrogen, (b) agriculture (%) in land cover, (c) water current velocity and (d) cobble (%).

**Author Contributions:** Conceptualization, M.-J.B. and Y.-S.P.; formal analysis, M.-J.B.; investigation, M.-J.B. and Y.-S.P.; writing—original draft preparation, M.-J.B.; writing—review and editing, Y.-S.P. All authors have read and agreed to the published version of the manuscript.

**Funding:** This work was supported by the Nakdonggang National Institute of Biological Resources (NNIBR), funded by the Ministry of Environment (MOE) of the Republic of Korea (grant number NNIBR201901111), and by the National Research Foundation of Korea (NRF) funded by the Korean government (MSIP, grant numbers NRF-2016R1A2B4011801 and NRF-2019R1A2C1087099).

**Acknowledgments:** We would like to thank the National Aquatic Ecological Monitoring Program (NAEMP) operated by the Ministry of Environment and National Institute of Environmental Research, Korea, for providing part of the dataset used.

**Conflicts of Interest:** The authors declare no conflicts of interest. The funders had no role in the design of the study; in the collection, analyses, or interpretation of data; in the writing of the manuscript, or in the decision to publish the results.

## References

1. Strong, E.E.; Gargominy, O.; Ponder, W.F.; Bouchet, P. Global diversity of gastropods (Gastropoda; Mollusca) in freshwater. In *Freshwater Animal Diversity Assessment. Developments in Hydrobiology*; Balian, E.V., Lévêque, C., Segers, H., Martens, K., Eds.; Springer: Dordrecht, The Netherlands, 2007; Volume 198, pp. 149–166.
2. Johnson, P.D.; Bogan, A.E.; Brown, K.M.; Burkhead, N.M.; Cordeiro, J.R.; Garner, J.T.; Hartfield, P.D.; Lepitzki, D.A.; Mackie, G.L.; Pip, E.; et al. Conservation status of freshwater gastropods of Canada and the United States. *Fisheries* **2013**, *38*, 247–282. [CrossRef]
3. Lopes-Lima, M.; Burlakova, L.E.; Karatayev, A.Y.; Mehler, K.; Seddon, M.; Sousa, R. Conservation of freshwater bivalves at the global scale: Diversity, threats and research needs. *Hydrobiologia* **2018**, *810*, 1–14. [CrossRef]
4. Bae, M.-J.; Kwon, Y.; Hwang, S.-J.; Chon, T.-S.; Yang, H.-J.; Kwak, I.-S.; Park, J.-H.; Ham, S.-A.; Park, Y.-S. Relationships between three major stream assemblages and their environmental factors in multiple spatial scales. *Ann. Limnol. Int. J. Lim.* **2011**, *47*, S91–S105. [CrossRef]
5. Williams, J.D.; Warren, M.L., Jr.; Cummings, K.S.; Harris, J.L.; Neves, R.J. Conservation status of freshwater mussels of the United States and Canada. *Fisheries* **1993**, *18*, 6–22. [CrossRef]
6. Abramovitzm, J.N. *Imperiled Waters Impoverished Future the Decline of Freshwater Ecosystems*; Worldwatch Institute: Washington, DC, USA, 1996; Volume Paper No. 128.
7. Clements, R.; Koh, L.P.; Lee, T.M.; Meier, R.; Li, D. Importance of reservoirs for the conservation of freshwater molluscs in a tropical urban landscape. *Biol. Conserv.* **2006**, *128*, 136–146. [CrossRef]

8.    Oehlmann, J.; Schulte-Oehlmann, U. Molluscs as bioindicators. In *Trace Metals and Other Contaminants in the Environment*; Markert, B.A., Breure, A.M., Zechmeister, H.G., Eds.; Elsevier: Amsterdam, The Netherlands, 2003; pp. 577–635.

9.    Waykar, B.; Deshmukh, G. Evaluation of bivalves as bioindicators of metal pollution in freshwater. *Bull. Environ. Contam. Toxicol.* **2012**, *88*, 48–53. [CrossRef] [PubMed]

10.   Heino, J.; Muotka, T.; Paavola, R.; Hämäläinen, H.; Koskenniemi, E. Correspondence between regional delineations and spatial patterns in macroinvertebrate assemblages of boreal headwater streams. *J. N. Am. Benthol. Soc.* **2002**, *21*, 397–413. [CrossRef]

11.   Richards, C.; Haro, R.; Johnson, L.; Host, G. Catchment and reach-scale properties as indicators of macroinvertebrate species traits. *Freshw. Biol.* **1997**, *37*, 219–230. [CrossRef]

12.   Townsend, C.R.; Doledec, S.; Scarsbrooks, M.R. Species traits in relation to temporal and spatial heterogeneity in streams: A test of habitat templet theory. *Freshw. Biol.* **1997**, *37*, 367–387. [CrossRef]

13.   Barmuta, L.A. Interaction between the effects of substratum, velocity and location on stream benthos: An experiment. *Aust. J. Mar. Freshw. Res.* **1990**, *41*, 557–573. [CrossRef]

14.   Downes, B.J.; Lake, P.S.; Schreiber, E.S.G. Habitat structure and invertebrate assemblages on stream stones: A multivariate view from the riffle. *Austral Ecol.* **1995**, *20*, 502–514. [CrossRef]

15.   Wohl, D.L.; Wallace, J.B.; Meyer, J.L. Benthic macroinvertebrate community structure, function and production with respect to habitat type, reach and drainage basin in the southern Appalachians. *Freshw. Biol.* **1995**, *34*, 447–464. [CrossRef]

16.   Poff, N.L. Landscape filters and species traits: Towards mechanistic understanding and prediction in stream ecology. *J. N. Am. Benthol. Soc.* **1997**, *16*, 19. [CrossRef]

17.   Strzelec, M.; Spyra, A.; Krodkiewska, M.; Serafinski, W. The long-term transformations of Gastropod communities in dam-reservoirs of Upper Silesia (Southern Poland). *Malacol. Bohemoslov.* **2005**, *4*, 41–47.

18.   Lewin, I.; Smoliński, A. Rare, threatened and alien species in the gastropod communities in the clay pit ponds in relation to the environmental factors (The Ciechanowska Upland, Central Poland). *Biodivers. Conserv.* **2006**, *15*, 3617–3635. [CrossRef]

19.   Greenwood, K.S.; Thorp, J.H. Aspects of ecology and conservation of sympatric, prosobranch snails in a large river. *Hydrobiologia* **2001**, *455*, 229–236. [CrossRef]

20.   Antoine, C.; Castella, E.; Castella-Müller, J.; Lachavanne, J.B. Habitat requirements of freshwater gastropod assemblages in a lake fringe wetland (Lake Neuchâtel, Switzerland). *Arch. Für Hydrobiol.* **2004**, *15*, 377–394. [CrossRef]

21.   Çabuk, Y.; Arslan, N.; Yılmaz, V. Species composition and seasonal variations of the Gastropoda in Upper Sakarya River System (Turkey) in relation to water quality. *Clean–Soilairwater* **2004**, *32*, 393–400.

22.   Li, F.; Chung, N.; Bae, M.-J.; Kwon, Y.-S.; Park, Y.-S. Relationships between stream macroinvertebrates and environmental variables at multiple spatial scales. *Freshw. Biol.* **2012**, *57*, 2107–2124. [CrossRef]

23.   Li, F.; Chung, N.; Bae, M.-J.; Kwon, Y.-S.; Kwon, T.-S.; Park, Y.-S. Temperature change and macroinvertebrate biodiversity: Assessments of organism vulnerability and potential distributions. *Clim. Chang.* **2013**, *119*, 421–434. [CrossRef]

24.   Kang, H.; Bae, M.-J.; Lee, D.-S.; Hwang, S.-J.; Moon, J.-S.; Park, Y.-S. Distribution patterns of the freshwater oligochaete *Limnodrilus hoffmeisteri* influenced by environmental factors in streams on a Korean nationwide scale. *Water* **2017**, *9*, 921. [CrossRef]

25.   Ministry of Environment/National Institute of Environmental Research. *The Survey and Evaluation of Aquatic Ecosystem Health in Korea*; The Ministry of Environment/National Institute of Environmental Research: Incheon, Korea, 2008; (In Korean with English Summary).

26.   Kwak, I.-S.; Lee, D.-S.; Hong, C.; Park, Y.-S. Distribution patterns of benthic macroinvertebrates in streams of Korea. *Korean J. Ecol. Environ.* **2018**, *51*, 60–70. [CrossRef]

27.   Lee, D.Y.; Lee, D.S.; Bae, M.J.; Hwang, S.J.; Noh, S.Y.; Moon, J.S.; Park, Y.S. Distribution patterns of odonate assemblages in relation to environmental variables in streams of South Korea. *Insects* **2018**, *9*, 152. [CrossRef] [PubMed]

28.   Bae, M.-J.; Li, F.; Kwon, Y.-S.; Chung, N.; Choi, H.; Hwang, S.-J.; Park, Y.-S. Concordance of diatom, macroinvertebrate and fish assemblages in streams at nested spatial scales: Implications for ecological integrity. *Ecol. Indic.* **2014**, *47*, 89–101. [CrossRef]

29. Ward, J.H. Hierarchical grouping to optimize an objective function. *J. Am. Stat. Assoc.* **1963**, *58*, 236–244. [CrossRef]

30. Oksanen, J.; Blanchet, F.G.; Kindt, R.; Legendre, P.; Minchin, P.; O'Hara, R.B.; Simpson, G.; Solymos, P.; Stevens, M.H.H.; Wagner, H. Vegan: Community Ecology Package. R Package Version. 2011. Available online: https://www.researchgate.net/publication/258996515_Vegan_Community_Ecology_Package_R_Package_Version_117-8 (accessed on 23 June 2018).

31. R Core Team. *R: A Language and Environment for Statistical Computing*; R Foundation for Statistical Computing: Vienna, Austria, 2017; Available online: https://www.R-project.org/ (accessed on 23 June 2018).

32. Dufrene, M.; Legendre, P. Species assemblages and indicator species: The need for a flexible asymmetrical approach. *Ecol. Monogr.* **1997**, *67*, 345. [CrossRef]

33. Roberts, D.W. Labdsv: Ordination and Multivariate Analysis for Ecology. R package version 1.8-0. 2016. Available online: https://CRAN.R-project.org/package=labdsv (accessed on 23 June 2018).

34. Oksanen, J.; Kindt, R.; Legendre, P.; O'Hara, B.; Simpson, G.L.; Stevens, M.H.H.; Wagner, H. Vegan: Community Ecology Package. R Package Version 1.13-1. 2008. Available online: http://vegan.r-forge.r-project.org/ (accessed on 23 June 2018).

35. Virtanen, R.; Luoto, M.; Rämä, T.; Mikkola, K.; Hjort, J.; Grytnes, J.A.; Birks, H.J.B. Recent vegetation changes at the high-latitude tree line ecotone are controlled by geomorphological disturbance, productivity and diversity. *Glob. Ecol. Biogeogr.* **2010**, *19*, 810–821. [CrossRef]

36. Breiman, L. Random forests. *Mach. Learn.* **2001**, *45*, 5–32. [CrossRef]

37. Cutler, D.R.; Edwards, T.C., Jr.; Beard, K.H.; Cutler, A.; Hess, K.T.; Gibson, J.; Lawler, J.J. Random forests for classification in ecology. *Ecology* **2007**, *88*, 2783–2792. [CrossRef]

38. Prasad, A.M.; Iverson, L.R.; Liaw, A. Newer classification and regression tree techniques: Bagging and random forests for ecological predictions. *Ecosystems* **2007**, *9*, 181–199. [CrossRef]

39. Liaw, A.; Wiener, M. Classification and Regression by randomForest. *R News* **2002**, *2/3*, 18–22.

40. Bae, M.-J.; Park, Y.-S. Diversity and distribution of endemic stream insects on a nationwide scale, South Korea: Conservation perspectives. *Water* **2017**, *9*, 833. [CrossRef]

41. Kwon, Y.-S.; Bae, M.-J.; Hwang, S.-J.; Kim, S.-H.; Park, Y.-S. Predicting potential impacts of climate change on freshwater fish in Korea. *Ecol. Inform.* **2015**, *29*, 156–165. [CrossRef]

42. Elith, J.; Leathwick, J.R.; Hastie, T. A working guide to boosted regression trees. *J. Anim. Ecol.* **2008**, *77*, 802–813. [CrossRef]

43. Stewart, P.M.; Butcher, J.T.; Swinford, T.O. Land use, habitat, and water quality effects on macroinvertebrate communities in three watersheds of a Lake Michigan associated marsh system. *Aquat. Ecosyst. Health Manag.* **2000**, *3*, 179–189. [CrossRef]

44. Karube, Z.I.; Okada, N.; Tayasu, I. Sulfur stable isotope signature identifies the source of reduced sulfur in benthic communities in macrophyte zones of Lake Biwa, Japan. *Limnology* **2012**, *13*, 269–280. [CrossRef]

45. Lodge, D.M.; Kelly, P. Habitat disturbance and the stability of freshwater gastropod populations. *Oecologia* **1985**, *68*, 111–117. [CrossRef]

46. Brown, K.M.; George, G.; Daniel, W. Urbanization and a threatened freshwater mussel: Evidence from landscape scale studies. *Hydrobiologia* **2010**, *655*, 189–196. [CrossRef]

47. Tchakonté, S.; Ajeagah, G.A.; Diomandé, D.; Camara, A.I.; Ngassam, P. Diversity, dynamic and ecology of freshwater snails related to environmental factors in urban and suburban streams in Douala-Cameroon (Central Africa). *Aquat. Ecol.* **2014**, *48*, 379–395. [CrossRef]

48. Camara, I.A.; Bony, Y.K.; Diomandé, D.; Edia, O.E.; Konan, F.K.; Kouassi, C.N.; Gourene, G.; Pointier, J.P. Freshwater snail distribution related to environmental factors in Banco National Park, an urban reserve in the Ivory Coast (West Africa). *Afr. Zool.* **2012**, *47*, 160–168. [CrossRef]

49. Evans, R.R.; Ray, S.J. Distribution and environmental influences on freshwater gastropods from lotic systems and springs in Pennsylvania, U.S.A., with conservation recommendations. *Am. Malacol. Bull.* **2010**, *28*, 135–150. [CrossRef]

50. Venter, O.; Brodeur, N.N.; Nemiroff, L.; Belland, B.; Dolinsek, I.J.; Grant, J.W. Threats to endangered species in Canada. *Aibs Bull.* **2006**, *56*, 903–910. [CrossRef]

51. Burlakova, L.E.; Karatayev, A.Y.; Karatayev, V.A.; May, M.E.; Bennett, D.L.; Cook, M.J. Endemic species: Contribution to community uniqueness, effect of habitat alteration, and conservation priorities. *Biol. Conserv.* **2011**, *144*, 155–165. [CrossRef]

52.  Matsuoka, K.; Kimura, T.; Kimura, S.; Yamaguchi, K.; Takayasu, K. Molluscan Fauna of the lower reaches of the Toyogawa River. *Sci. Rep. Toyohashi Mus. Nat. Hist.* **1999**, *9*, 15–24, (In Japanese with English Abstract).

53.  Lowe, S.; Browne, M.; Boudjelas, S. *100 of the World's Worst Invasive Alien Species: A Selection from the Global Invasive Species Database*; Invasive Species Specialist Group: Auckland, New Zealand, 2000.

54.  Bae, M.-J.; Kwon, Y.-S.; Park, Y.-S. Effects of global warming on the distribution of overwintering *Pomacea canaliculata* (Gastropoda: Ampullariidae) in Korea. *Korean J. Limnol.* **2012**, *45*, 453–458.

55.  Lee, D.-S.; Park, Y.-S. Factors affecting distribution and dispersal of *Pomacea canaliculata* in South Korea. *Korean J. Ecol. Environ.* **2020**, *53*, 185–194. [CrossRef]

56.  Bae, M.-J.; Chon, T.-S.; Park, Y.-S. Modeling behavior control of golden apple snails at different temperatures. *Ecol. Model.* **2015**, *306*, 86–94. [CrossRef]

57.  Seuffert, M.E.; Martín, P.R. Thermal limits for the establishment and growth of populations of the invasive apple snail Pomacea canaliculata. *Biol. Invasions* **2017**, *19*, 1169–1180. [CrossRef]

58.  Carlsson, N.O.L.; Brönmark, C.; Hansson, L.-A. Invading herbivory: The golden apple snail alters ecosystem functioning in Asian wetlands. *Ecology* **2004**, *85*, 1575–1580. [CrossRef]

59.  Clavero, M.; Brotons, L.; Pons, P.; Sol, D. Prominent role of invasive species in avian biodiversity loss. *Biol. Conserv.* **2009**, *142*, 2043–2049. [CrossRef]

 *water*

*Review*

# Habitats and Diversity of Subterranean Macroscopic Freshwater Invertebrates: Main Gaps and Future Trends

**Elzbieta Dumnicka [1,*], Tanja Pipan [2]**  **and David C. Culver [3]**

1    Institute of Nature Conservation, Polish Academy of Sciences, 31-120 Kraków, Poland
2    Karst Research Institute, Research Centre of the Slovenian Academy of Sciences and Arts, 1000 Ljubljana, Slovenia; tanja.pipan@zrc-sazu.si
3    Department of Environmental Science, American University, Washington, DC 20016, USA; dculver@american.edu
*    Correspondence: dumnicka@iop.krakow.pl

Received: 15 June 2020; Accepted: 24 July 2020; Published: 31 July 2020

**Abstract:** Caves are the best studied aquatic subterranean habitat, but there is a wide variety of these habitats, ranging in depth below the surface and size of the spaces (pore or habitat size). Both factors are important in setting limits to species composition and richness. In addition to caves, among the most important shallow aquatic subterranean habitats are the hyporheal (underflow of rivers and streams), the hypotelminorheal (very superficial drainages with water exiting in seeps), epikarst, and calcrete aquifers. Although it is little studied, both body size and species composition in the different habitats is different. Because of high levels of endemism and difficulty in access, no subterranean habitats are well sampled, even caves. However, there are enough data for robust generalizations about some geographic patterns. Individual hotspot caves are concentrated in the Dinaric region of southern Europe, and overall, tropical regions have fewer obligate aquatic cave dwellers (stygobionts). In all subterranean aquatic habitats, regional diversity is much higher than local diversity, but local diversity (especially single cave diversity) may be a useful predictor of regional species richness. In Europe there is a ridge of high aquatic subterranean species richness basically extending east from the French–Spanish border. Its cause may be either high productivity or that long-term temperature oscillations are at a minimum. With increased collecting and analysis, global and continental trends should become clearer.

**Keywords:** calcrete aquifer; epikarst; hyporheal; hypotelminorheal; stygobiont

## 1. Introduction

The existence of eyeless and depigmented animals in the darkness of cave streams has been known since at least 1537 [1]. Of course, caves themselves have been known since at least the Paleolithic [2]. Groundwater also has a long history of human knowledge and use, and the first known well dates back to the Neolithic [3]. On the biological side, in 1907 Racovitza [4] mentioned studies from the 1890s on eyeless, depigmented species from artesian wells in Texas and wells in the Canterbury Plain of New Zealand.

It is probably Racovitza [4] who first pointed to the potential advantages of considering all terrestrial and freshwater aphotic habitats and their inhabitants together. Earlier, North American neo-Lamarckians had a strong interest in the evolution of eyelessness due to its connection to evolution by disuse (e.g., Packard [5]). While there was a general recognition of the unity of the subterranean domain, especially by European scientists, e.g., Ginet and Decou [6], it was not until the publication of the groundbreaking book *"Groundwater Ecology"* [7] that at least the aquatic side of subterranean biology became a distinct discipline with its own tenets.

In this review, we consider three areas. The first is the range of aquatic subterranean habitats and what unites and divides them. There have been two overarching classifications of freshwater subterranean habitats. One, due to Botosaneanu [8] who divided aquatic subterranean habitats (he included the marine interstitial) into two branches—"milieu perméables en grande" and "milieu perméables en petit". He included more than 30 habitats, and the two most familiar are caves and the hyporheic, the underflow of streams and rivers. The other is due to Culver and Pipan [9,10] in which they attach equal, if not greater, weight to the vertical depth of the habitat compared to habitat size. We reassess these divisions and revisit the range of aquatic subterranean habitats.

The second area we review is the vexing question: how complete is our knowledge of the subterranean biota? In addition to the nearly universal shortfall of taxonomists to describe new species that have already been collected (the Linnean impediment), there are regions of the Earth's surface that have been little sampled for aquatic subterranean fauna, especially in the tropics (the Wallacean impediment), and there are subterranean habitats that are poorly sampled due to sampling difficulties (the Racovitzan impediment [11]). All of these problems are especially severe in aquatic subterranean habitats because of the high levels of endemism of the species [12,13]. What can we say about overall fauna patterns in light of this uncertainty?

The third area we review is: what is the global pattern of richness of aquatic subterranean invertebrates? After reviewing the pattern of individual hotspot sites globally, we review some of the potential "fixes" to incomplete data, especially functions such as those of Chao [14] that estimate missing species, as well as assessing the potential impact of missing data (e.g., Zagmajster et al. [15]).

## 2. Aquatic Subterranean Habitats

Botosaneanu's [8] compilation of subterranean habitats, produced as an addendum to his extensive review of the subterranean fauna is an appropriate place to begin (Table 1). We separated out from his dichotomy of "perméable en grand" and "perméable en petit" the categories of springs, which he included in "perméable en petit". Springs are both habitats themselves (ecotones between surface and subterranean waters), but also convenient collecting points for subterranean species from more inaccessible (often unknown) subterranean habitats. These habitats may have fine or coarse sediments and may be shallow or deep.

Botosaneanu suggested only species limited to subterranean habitats be considered, but in practice it makes more sense to consider all species in springs (e.g., [16,17]), especially since not only species limited to subterranean habitats are blind and eyeless [18]. Springs themselves can be classified in a number of different ways, the oldest being based on discharge rate [19]. Other classifications, which can be quite elaborate (see Springer and Stevens [20]), are based on characteristics of the hydraulic head, geomorphologic structure, and water quality and temperature [21]. All of these are more elaborate than Botosaneanu's [8] (Table 1). He listed springs as porous habitats [8] but we give them a separate category. Perhaps the reason that speleobiologists have not taken up more elaborate subdivisions of springs is that the fauna of springs often has few, if any, species showing the characteristics of subterranean life such as reduced eyes and pigment (see Botosaneanu [22]). However, this is not always the case. Dumnicka and Galas [23] show that a significant fraction of the subterranean fauna of Poland can be found in springs. Certainly, the classification of springs from the point of view of the groundwater fauna needs more attention. For example, Dumnicka et al. [17] show that substrate in the spring has a major effect on faunal composition.

**Table 1.** Aquatic subterranean habitats: after Botosaneanu [8]—simplified and modified. The shallow/deep dichotomy is based on a diving line at 10 m [10].

| Major Categories | Botosaneanu's [8] Divisions, Supplemented | Pore Size (after [8]) | Corrections to Pore Size | Depth | Replicate of Surface Habitat |
|---|---|---|---|---|---|
| Large habitats (caves in karst or pseudokarst) | cave water in general | large | | variable | variable |
| | percolation water (rimstone pools) | large | small [1] | shallow | No |
| | epiphreatic (streams) | large | | variable | Yes |
| | phreatic lakes | large | | deep | Yes |
| | cenotes | large | | deep | Yes |
| | anchialine | large | | variable | No |
| | lava tubes, mines, etc. | large | | shallow | Yes |
| | artesian wells | large | | deep | No |
| | calcrete aquifers [2] | | large | variable | No |
| Porous habitats | Alluvial wells | small | | deep | No |
| | hyporheal | small | | shallow | No |
| | hyporheal in caves | small | | variable | No |
| | Water on border of fw stagnant water | small | | shallow | No |
| | hypotelminorheal | small | | shallow | No |
| | artificial filters of sand or gravel | small | | shallow | No |
| | interstitial water of marine beaches | small | | shallow | No |
| | interstitial water of marine sublitoral | small | | shallow | No |
| | interstitial of brackish or hypersaline water bodies | small | | shallow | No |
| Ecotones (springs) | general | small | | variable | n/a |
| | karst | small | | variable | n/a |
| | phreatic | small | | variable | n/a |
| | hypotelminorheic | small | | shallow | n/a |
| | thermal springs | small | | variable | n/a |
| | travertine springs [3] | | small | shallow | n/a |

[1]—in epikarst the pores are often small with miniature cavities [24]; [2]—this habitat is typical for Australia—see [25,26]; [3]—described by Pentecost [27].

Botosaneanu's [8] division of cave habitats vertically (percolating (epikarst), vadose, epiphreatic, phreatic) follows the convention of hydrogeology. A similar classification, based on Leruth [28], was used by Howarth and Moldovan [29]. They identified five aquatic cave habitats:

1. Highly dynamic flowing waters (sinking streams);
2. Slow-flowing waters and lakes;
3. Gours or pools formed on flowstone;
4. Small pools on clay or mud;
5. Dripping or percolation water.

There can be more elaborate subdivisions. For example, Poulson [30] distinguishes several types of cave streams, such as shallow streams and moderately deep master shaft drain streams. While no doubt useful in the context of Mammoth Cave, where he worked, their generality seems very limited. Each of the cave subdivisions has its own strengths and weaknesses, and it seems counterproductive to a priori decide how detailed such a subdivision should be.

Small pore habitats, categorized by Botosaneanu [8] (Table 1), were largely divided on the basis of water flow and salinity. An interesting feature of the classification is the hypotelminorheal—a persistent wet spot, a kind of perched aquifer, fed by subsurface water in a slight depression in an area of low to moderate slope, rich in organic matter, underlain by an impermeable clay layer typically less than 50 cm below the surface—see Culver and Pipan [9] and Meštrov [31]. It appears in the classification both as a porous habitat and as a spring (called seepage springs by Keany et al. [32]). If more were known about the drainage area of all springs, this duality of classification should occur for all spring types. The fauna collected at seepage springs seems to be a mixture of species primarily found in the hypotelminorheal and species primarily found in the seepage spring itself [33].

Just as caves dominate large pore (diameter) habitats, the hyporheal dominates small pore (diameter) habitats, at least in terms of the amount of research done. Orghidan [34], (see Käser [35] for an English translation) coined the term and defined it as the zone of interstitial spaces constituted by the sediments of the stream bed. To our knowledge there have not been formally proposed subdivisions but even at the scale of a few meters there are differences, especially due to upwelling and downwelling [36]. In different streams, the connections between the hyporheic, the groundwater zone, and underlying impermeable substratum vary and have a profound effect on the composition and abundance of the fauna. Malard et al. [36] list five cases but do not formally name them:

1. No hyporheic zone.
2. Hyporheic zone created only by advected channel water (no ground water).
3. Hyporheic zone created by advection by both channel water and ground water.
4. Hyporheic zone created only by infiltration of channel water beneath the stream bed (no parafluvial flow).
5. A perched hyporheic zone created only by infiltration of channel water beneath the stream bed.

Their fifth category would appear to include seepage springs (outlet of the hypotelminorheic).

As knowledge of subterranean habitats has grown, the dichotomy between large and small pore habitats has grown increasingly problematic. Habitats such as calcrete aquifers and the hypotelminorheic are likely intermediate in pore size, at least based on the size of the organisms found in these habitats. Although epikarst and percolating water are part of karst, they are small pore habitats, based on the size of the inhabitants [10]. There is accumulating evidence [24] that the size of the habitat spaces has an impact on body size. Their model of how this works is shown in Figure 1 and the overall pattern in relationship to habitat categories is shown in Figure 2.

Given these size differences in both habitat and organism, one would expect the community composition in different habitats to be quite different. Dumnicka et al. [37], basing on 280 records for interstitial habitat, 150 for cave waters and 50 for wells, looked at this question for the Polish groundwater annelids fauna. Somewhat surprisingly, relatively little separation of habitats was found, although there was a tendency for the interstitial fauna to differ from the cave fauna along the first axis of their correspondence analysis (Figure 3) (Supplementary Materials). The well fauna, collected in various kinds of wells, overlapped broadly with both. Hahn and Fuchs [38] found a similar pattern for the German fauna. It is curious that there have not been more analyses of this question, and it is worth pursuing.

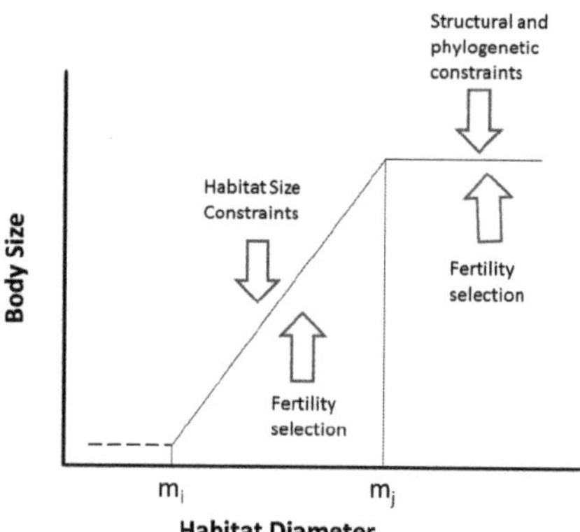


**Figure 1.** Hypothetical relationship between pore size (habitat diameter) in subterranean habitats and body size, with selective forces indicated by arrows. Below a minimum ($m_i$), there is not sufficient space for animals to occur without burrowing. Above a maximum ($m_j$), body size is likely constrained by other factors, such as phylogenetic and structural constraints. The relationship need not be linear but is presented as such for simplicity. From Pipan and Culver [24]. Used with permission of the National Speleological Society (www.caves.org).

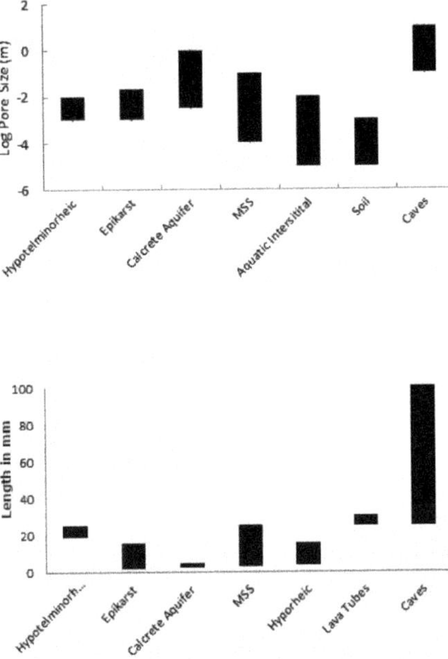

**Figure 2.** Histograms of body lengths of inhabitants (bottom) and log of pore size (top) for different subterranean habitats. From Pipan and Culver [24], modified. Log pore (habitat) size was used because the range of habitat sizes ranged over several magnitudes. The MSS (milieu souterrain superficiel) is the habitat of interconnected cracks and crevices of scree slopes, especially covered ones. Used with permission of the National Speleological Society (www.caves.org).

An interesting question is how many species are found in the different subterranean habitats in a region. In Poland, which was largely glaciated, Dumnicka and Galas [23] demonstrate that there are more records per sample in interstitial waters and wells than in caves. In fact, there are more records per sample in springs than in caves (Figure 4). As far as we can determine, these are the only data of this type available. It would be interesting to compare the Polish data with data from an unglaciated area, areas which generally have a much richer cave fauna.

Culver and Pipan [10,39], Pipan and Culver [18] and Blatnik et al. [40] suggested that there was a third major category of subterranean habitats, close to the surface and with intermediate-sized habitat spaces—shallow subterranean habitats (see Table 1). In their book-length treatment of the topic, Culver and Pipan [10] emphasized the vertical division—the distance from the surface and defined shallow subterranean habitats as occurring less than 10 m from the surface. They suggested several unifying features:

- Absence of light;
- Close surface-subsurface connections (except for calcrete aquifers);
- Availability of organic carbon and nutrients;
- Generally small habitat (pore) size.

Habitats include the hypotelminorheal, epikarst, hyporheal, and calcrete aquifers. Halse [25] calls calcrete aquifers deep subterranean habitats but gives no criteria for this choice. Calcretes in the Pilbara region are deeper than 10 m [13] while those in the Yilgarn are often less than 10 m [41]. Of course, all caves with natural entrances occur in part at depths of less than 10 m. Culver and Pipan [10] included only habitats that were typically less than 10 m in depth. Whether calcretes are included or not, the grouping of habitats into shallow versus deep has proven to be frequently used, although it remains to be seen how useful the distinction is.

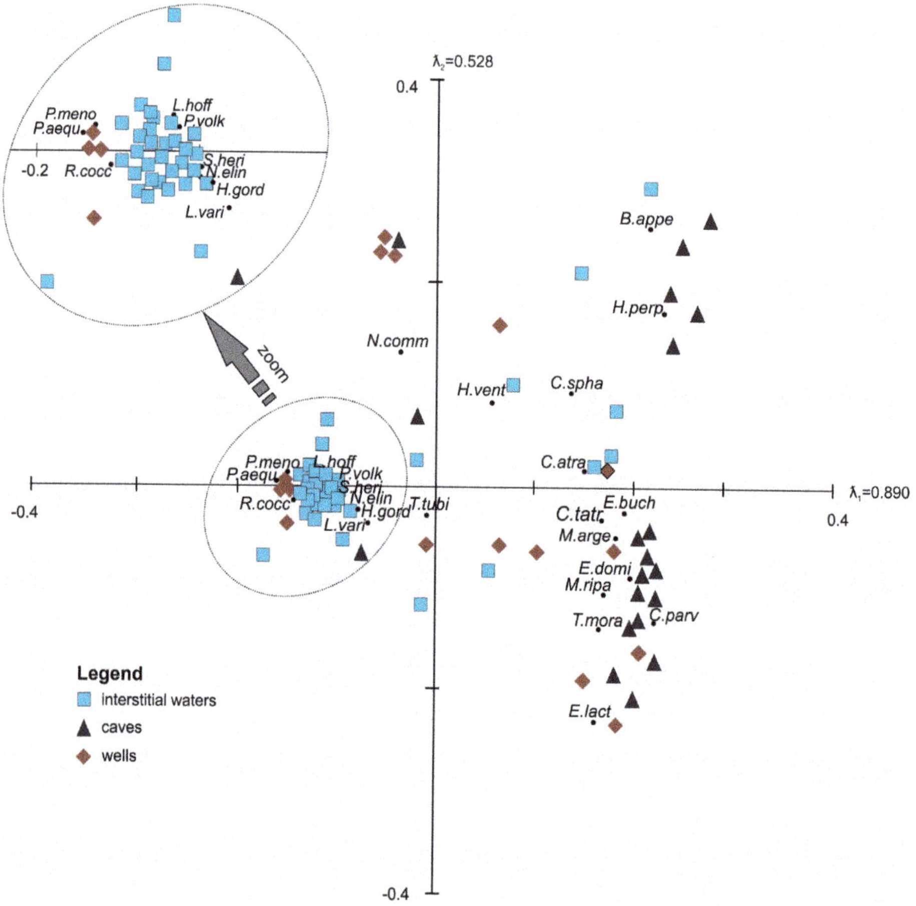

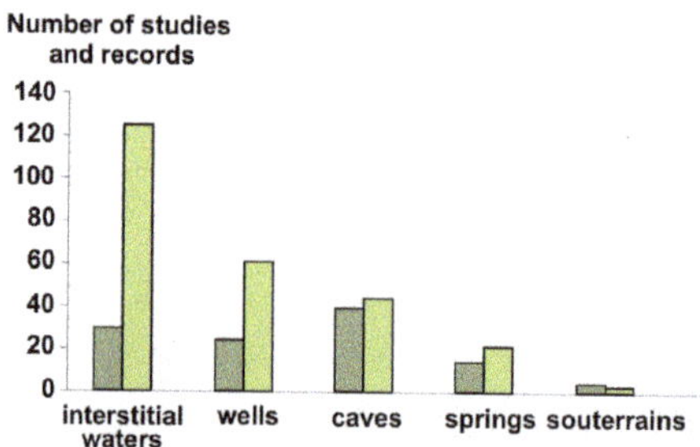

**Figure 3.** Correspondence analysis (CA) ordination diagram with both samples and species plotted. Abbreviations of species names are explained in Table S1. From Dumnicka et al. [37], modified.



**Figure 4.** Number of studies performed in particular habitats (green bar) and number of records of specialized subterranean species found in them (light green bar) in Poland. From Dumnicka and Galas [23], modified.

## 3. The Struggle to Measure Aquatic Subterranean Biodiversity

Aquatic subterranean biodiversity surveys share with most other invertebrate surveys both a Linnean shortfall (not enough taxonomists) and a Wallacean shortfall (not all areas sampled). However, there are a number of taxonomists, especially in Europe, actively working on major subterranean groups, including amphipods, copepods, isopods, and oligochaetes. With respect to the Wallacean shortfall, tropical caves, if not other subterranean habitats, are being increasingly well sampled (e.g., [42–44]).

However, there are a number of aspects of the subterranean fauna that make unbiased sampling very difficult. Nearly all the analyses and datasets are based on those species limited to subterranean habitats—stygobionts for aquatic habitats. However, in most subterranean habitats there are species that show little morphological modification for subterranean life and/or are also found in surface habitats [18]. Such stygophiles are part of the subterranean community and reproduce in subterranean habitats. Analysis of stygobionts is an analysis of the highly evolved component of subterranean communities specialized and limited to subterranean habitats, not the entire subterranean community.

A second issue is the omnipresence of undescribed species. Some of these species that have been collected are awaiting taxonomic analysis, some have been observed but not collected, and some have not even been observed but are considered likely to be present in a region by expert taxonomists. Obviously the last two are fraught with the likelihood of overestimation or exaggeration of numbers [45]. Bolded letters spell out PASCALIS and it is usually written this way. Collected, undescribed species present a special problem. If they are ignored, then some regions, like Brazil—see Trajano and Bichuette [46]—will appear to be species-poor when in fact they are species rich. One possible solution to the problem is to apply a "discount rate". For example, Culver et al. [45] report that of 19 species listed as undescribed by Holsinger et al. [47] in their enumeration of the West Virginia cave fauna, six turned out to be previously described species. This results in a discount rate of 0.68. The omnipresence of undescribed species can also lead to inappropriate comparisons between regions based on data taken from different time frames—what Culver et al. [45] term the fallacy of provincialism.

A third and related issue is the high frequency of local endemism, often single site endemism. An example of this is from the large-scale European study of subterranean biodiversity, PASCALIS (Protocols for the Assessment and Conservation of Aquatic Life in the Subsurface) [48]. In this study of all known stygobionts from six European countries, there was a strong negative relationship between number of species and number of $0.2° \times 0.2°$ grid cells occupied by a species (Figure 5). Of the 930 described stygobionts, 396 were known from a single grid cell.

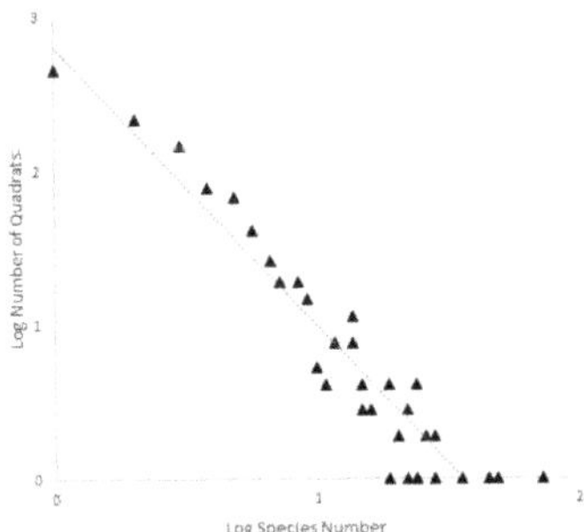


**Figure 5.** Relationship between the log of the numbers of quadrats and log of the numbers of species for the PASCALIS project. The relationship between the two is Y (number of quadrats) = 677 × X (Number of Species)$^{-1.76}$. Data from Deharveng et al. [49].

This means that unless all cells are sampled, many single cell endemics will be missed. In practice it appears that the pattern of species richness is unaffected if single cell and other narrow endemics are not included [50]. More generally, missing species do not seem to affect geographic pattern. Deharveng et al. [49] measured these missing species using a jackknife procedure that resamples and takes into account the number of narrow endemics. Their results, by country, are shown in Table 2.

**Table 2.** Groundwater biodiversity in six European countries—from Deharveng et al. [49], modified.

| Country | No. Sampled Cells | No. Sampled Sites | No. Observed Species | No. Predicted ssp. (Jackknife) |
| --- | --- | --- | --- | --- |
| Belgium | 17 | 155 | 33 | 43 |
| France | 566 | 1712 | 320 | 434 |
| Italy | 337 | 1580 | 288 | 394 |
| Portugal | 24 | 34 | 48 | 88 |
| Slovenia | 54 | 491 | 183 | 246 |
| Spain | 241 | 737 | 216 | 308 |
| ALL | 1228 | 4709 | 930 | 1291 |

The final impediment is the Racovitzan impediment, the incompleteness of sampling of subterranean habitats [11]. Caves are incompletely sampled because new caves are constantly being discovered and the number of known caves is very large (e.g., more than 10,000 in Slovenia). While sampling all caves is neither possible nor necessary, the high levels of endemism makes thorough sampling important. Nevertheless, there are some very large datasets for subterranean animals. Probably the largest is that of Zagmajster et al. [51], who assembled data for 21,000 occurrences of 1570 European aquatic subterranean species. The situation for non-cave aquatic subterranean species is particularly difficult because neither epikarst, the hyporheal, or the hypotelminorheal can be sampled easily. For the hyporheal, most samples are taken by pumping water out of the habitat through a fine mesh filter, and then sorting the samples. The Bou–Rouch pump [52,53] in many ways made the sampling of hyporheal possible. Prior to the development of a continuous filtering device [54], epikarst could only be sampled very indirectly by sampling drip pools, themselves biased samples of the organisms in dripping water [55]. The situation is even more dire for the hypotelminorheal. No good sampling device exists, and the habitat must be destructively sampled. However, Niemiller et al. [56] were able to show that eDNA of seep amphipods (genus *Stygobromus*) could be detected, allowing for the possibility of non-destructive sampling.

There is one way that at least the extent of undersampling can be better understood and that is to report on sites where no specimens were found. There has been a reluctance to do this, perhaps based on the sense that empty samples are failed samples. However, such reporting can be extremely informative. Dumnicka et al. [57] report those quadrats in Poland where they failed to find any water mites specialized for subterranean life, and almost all of these sites were in the glaciated areas of Poland, where the specialized subterranean fauna should be rare (Figure 6).

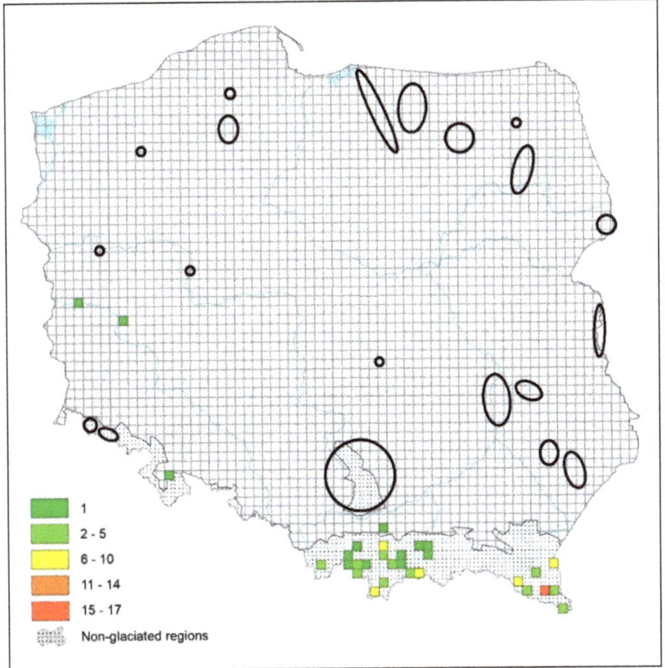



**Figure 6.** The number of stygobiotic water mites recorded in various squares in Poland. Ellipses (or circles) represent areas in which water mite fauna was studied in various types of surface/subterranean waters but stygobionts were not found. From Dumnicka et al. [57], modified.

## 4. Geographic Patterns of Species Richness in Aquatic Subterranean Invertebrates

The easiest, but certainly not the best, measure of species richness is the species richness at a single subterranean site, $\alpha$-diversity. It is an inadequate measure because $\alpha$- diversity is a small fraction of the species richness of a region, $\gamma$-diversity [58]. However, the data are much easier to accumulate since the regional analysis requires data on many sites [59]. Culver and Sket [60] published the first list, which included 20 caves and wells with 20 or more obligate subterranean species, including both aquatic and terrestrial species. Since then, the number of such sites has at least doubled, and most recently Culver and Pipan [61] published a list of sites with 25 or more stygobiotic species (Table 3).

**Table 3.** Caves and other karst sites with 25 or more obligate aquatic subterranean species (stygobionts). From Culver and Pipan [61], modified.

| Site Name | Country | No. of Species |
| --- | --- | --- |
| Postojna–Planina Cave System | Slovenia | 48 |
| Vjetrenica | Bosnia and Hercegovina | 40 |
| Walsingham Cave | Bermuda | 37 |
| Triadou wells | France | 34 |
| Robe River | Australia | 32 |
| Jameos del Aqua | Lanzarote, Canary Islands, Spain | 32 |
| Križna Jama | Slovenia | 29 |
| Logarček | Slovenia | 28 |
| Edwards Aquifer | Texas, USA | 27 |

Several points emerge. First, with the exception of the Robe River in western Australia, none of the sites are in the tropics or sub-tropics. Second, there is a concentration of hotspots in the Dinaric karst, with four of ten sites in this region.

Counts of numbers of species by country (corrected for size) give a similar picture to Table 3. There is a broad band of high species richness in the north temperate zone (China is relatively unknown) as well as in Australia (Figure 7), where aquatic subterranean diversity is largely found in calcrete aquifers, accessible only by wells [25,26].

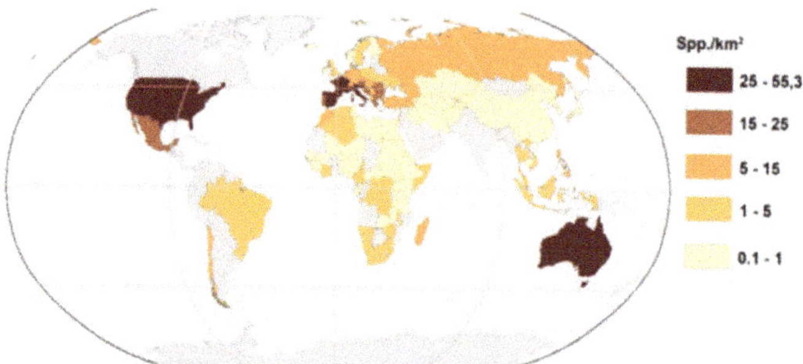


**Figure 7.** Distribution of obligate subterranean species by country, corrected for area. See Zagmajster et al. [62] for details. With permission of Springer Nature.

Calcrete aquifers are carbonate deposits that form in the vicinity of the water table as a result of evaporation of groundwater—see Culver and Pipan [10]. Aquatic subterranean species richness is very high in both the Yilgarn and Pilbara aquifers of western Australia, but it remains difficult to compare with other areas because of the distinctive and different way researchers have measured species richness in Australia [45,63].

More detailed analyses at the global scale are not available, but there is more detailed information available for Europe [48]. Zagmajster et al. [51] showed that there is a ridge of high species richness in southern Europe (Figure 8), one that corresponds to a similar ridge in terrestrial species richness [59]. The explanation for this ridge may be that it corresponds to a ridge of primary productivity or that long term temperature oscillations are at a minimum.

The above analyses of the patterns of aquatic subterranean species richness is an analysis of only the specialists for cave life. e.g., stygobionts. Many cave streams, especially in glaciated areas, have functioning, reproducing communities but without any specialized species. A good example of this is the aquatic fauna in Swildon's Hole, a nearly 10 km long cave in the Mendip Hills of the United Kingdom. Knight [64] found 38 taxa in the main stream, mainly Trichoptera and Diptera. There were no stygobionts, yet a functioning community. The situation in the tributaries was different where three of ten species were specialized for subterranean life.

Very little is known about the geographic pattern of species richness in non-cave aquatic subterranean habitats. The PASCALIS project [48] included extensive sampling of the hyporheal, but only eight relatively small regions were included in the study (Figure 8): Wallonia (Belgium), Jura (France), Rousillon (France), Cantabria (Spain), Padano–Alpine region (Italy), Slovenia, Rhône valley (France), and Garonne (France). Malard et al. [58] showed that species richness was highest in Slovenia, followed by the Rhone and the Garonne (Figure 9). Slovenia and the Garonne are on the ridge of aquatic species richness (Figure 8). One of the features of porous aquifers in general and hyporheal in particular is its fine scale heterogeneity, which is evident in Figure 9.

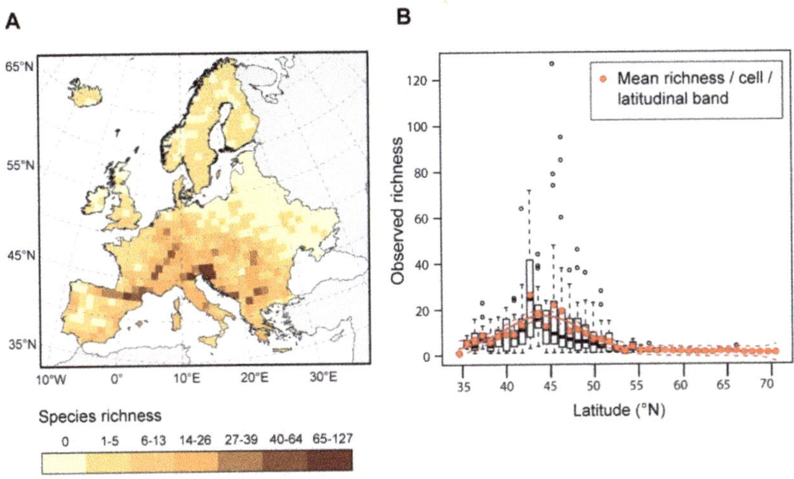

**Figure 8.** Map of species richness patterns of European stygobionts. (**A**) Species richness of 10,000 km$^2$ cells. (**B**) Relationship between the cell average of species richness per 0.09′ latitudinal band and latitude. Black horizontal bars and boxes show the median and interquartile range, respectively, for latitudinal banks. The maximum length of each whisker is 1.5 times the interquartile range and open circles represent outliers. The thick red line is the fit of generalized additive model to the averages of latitudinal bands. From Zagmajster et al. [51]. With permission of Springer Nature.

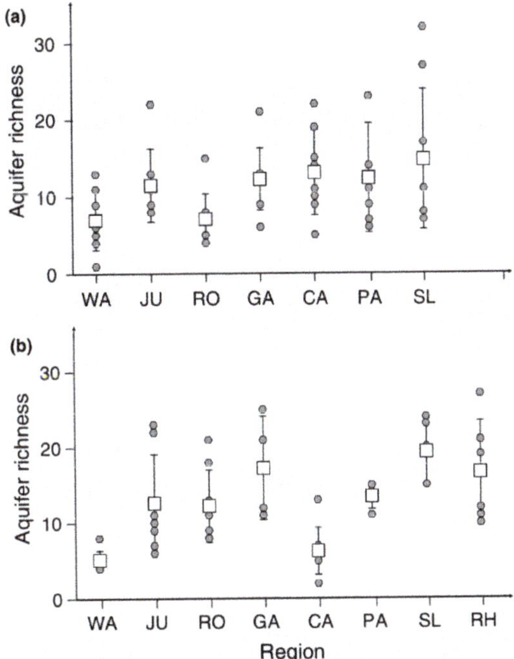

**Figure 9.** Differences in species richness (+SD) for porous aquifers from Wallonia (WA), Jura (JU), Roussilon (RO), Cantabria (CA), Padano–Alpine region (PA), Slovenia (SL), Rhône River valley (RH) and Garonne (GA) regions. (**a**)—karst, (**b**)—interstitial. From Malard et al. [58], modified.

Pipan et al. [65] did a small regional study of epikarst copepods in Slovenia. They analyzed 81 drips from13 caves in three karst areas of Slovenia. As with Malard et al.'s analysis, they found small scale differences and that α diversity (within drip species richness) was small compared to β diversity. Of the 18 species, only three were accounted for by α diversity, three by differences within a cave, six by differences between caves in a region, and 18 by differences among regions (Figure 10). When the data are viewed in another way, one that emphasizes the occurrence of "hotspot" drips, a different pattern emerges.

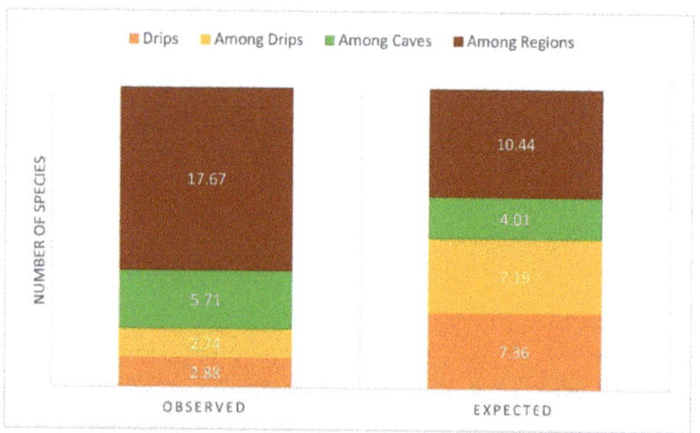


**Figure 10.** Relative contribution of within drip (α-diversity), among drip, among cave, and among region diversity (all β-diversity) to the overall diversity of 30 epikarst copepod species, relative to random expectation. From Pipan et al. [65].

A few drips contribute a disproportionate share of species diversity. The most species-rich drip in the Dinaric karst has 10 species and the entire Županova jama has 13 species, so this drip contributes 40 percent of the species diversity known from the entire Dinaric karst! The task of assessing epikarst species diversity would be considerably simplified if we had a method of determining which drips were hotspots prior to sampling, but we do not.

## 5. Conclusions

Obligate aquatic subterranean organisms (stygobionts) occur in a variety of subterranean habitats that vary both in depth and in pore size. This fauna is relatively well known in Europe and North America, and poorly known elsewhere. Sampling this fauna, particularly the non-cave fauna remains difficult, resulting in a Racovitzan shortfall for these habitats. Species richness is highest in mid-temperate latitudes and the Dinaric karst in southern Europe is a hotspot of stygobiotic species richness. Endemism is high and consequently β diversity is much higher than α diversity.

With increased amounts of data and new analytical tools, both continental and global patterns of species richness, and its explanation, should become clearer.

**Supplementary Materials:** The following are available online at http://www.mdpi.com/2073-4441/12/8/2170/s1, Table S1: Full names of species cited in Figure 3 and the text.

**Author Contributions:** Conceptualization, E.D.; methodology, D.C.C., E.D., T.P.; validation, D.C.C., E.D., T.P.; formal analysis, D.C.C., E.D., T.P.; investigation, D.C.C., E.D., T.P.; resources, E.D.; data curation, D.C.C., E.D., T.P.; writing—original draft preparation, D.C.C.; writing—review and editing, D.C.C., E.D., T.P.; visualization, E.D., T.P.; supervision, E.D.; project administration, D.C.C., E.D., T.P. All authors have read and agreed to the published version of the manuscript.

**Funding:** This research received no external funding.

**Conflicts of Interest:** The authors declare no conflict of interest.

## References


1. Romero, A. *Cave Biology: Life in Darkness*; Cambridge University Press: Cambridge, UK, 2009; ISBN 978-0-521-82846-8.
2. Shaw, T. *History of Cave Science: The Exploration and Study of Limestone Caves, to 1900*, 2nd ed.; Sydney Speleological Society: Sydney, Australia, 1992; ISBN 0-646-12503-6.
3. Tegel, W.; Elburg, R.; Hakelberg, D.; Stäuble, H.; Büntgen, U. Early Neolithic water wells reveal the world's oldest wood architecture. *PLoS ONE* **2012**, *7*, e51374. [CrossRef] [PubMed]
4. Racovitza, E.G. Essai sur les problems biospéologiques. *Arch. Zool. Exp. Gén. IV* **1907**, *6*, 371–488.
5. Packard, A.S. The cave fauna of North America, with remarks on the anatomy of brain and the origin of the blind species. *Mem. Nat. Acad. Sci. USA* **1888**, *4*, 1–156.
6. Ginet, R.; Decou, V. *Initiation à la Biologie et à L'écologie Souterraines*; Delarge, J-P.: Paris, France, 1977; ISBN 2-7113-0073-0.
7. Gibert, J.; Danielopol, D.L.; Stanford, J.A. (Eds.) *Groundwater Ecology*; Academic Press: San Diego, CA, USA, 1994; ISBN 978-0-08-050762-0.
8. Botosaneanu, L. (Ed.) *Stygofauna Mundi*; EJ Brill: Leiden, The Netherlands, 1986; ISBN 90-04-075712.
9. Culver, D.C.; Pipan, T. Redefining the extent of the aquatic subterranean biotope-shallow subterranean habitats. *Ecohydrolgy* **2011**, *4*, 721–730. [CrossRef]
10. Culver, D.C.; Pipan, T. *Shallow Subterranean Habitats. Ecology, Evolution, and Conservation*; Oxford University Press: Oxford, UK, 2014; ISBN 978-0-19-964617-3.
11. Ficetola, G.F.; Canedoli, C.; Stoch, F. The Racovitzan impediment and the hidden biodiversity of unexplored environments. *Conserv. Biol.* **2019**, *33*, 214–216. [CrossRef]
12. Trontelj, P.; Douady, C.J.; Fišer, C.; Gibert, J.; Gorički, Š.; Lefebure, T.B.; Zakšek, V. A molecular test for cryptic diversity in ground water: How large are the ranges of macro-stygobionts? *Freshw. Biol.* **2009**, *54*, 727–744. [CrossRef]
13. Eberhard, S.M.; Halse, S.A.; Williams, M.R.; Scanlon, M.D.; Cocking, J.; Barron, H.J. Exploring the relationship between sampling efficiency and short-range endemism for groundwater fauna in the Pilbara region. *Freshw. Biol.* **2009**, *54*, 885–901. [CrossRef]
14. Chao, A. Estimating the population size for capture-recapture data with unequal catchability. *Biometrics* **1987**, *43*, 783–791. [CrossRef]
15. Zagmajster, M.; Culver, D.C.; Christman, M.C.; Sket, B. Evaluating the sampling bias in pattern of subterranean species richness: Combining approaches. *Biodivers. Conserv.* **2010**, *19*, 3035–3048. [CrossRef]
16. Dumnicka, E. Stygofauna associated with springfauna in southern Poland. *Subterr. Biol.* **2005**, *3*, 29–36.
17. Dumnicka, E.; Galas, J.; Koperski, P. Benthic invertebrates in karst springs: Does substratum or location define communities? *Int. Rev. Hydrobiol.* **2007**, *92*, 452–464. [CrossRef]
18. Pipan, T.; Culver, D.C. Convergence and divergence in the subterranean realm: A reassessment. *Biol. J. Linn. Soc.* **2012**, *107*, 1–14. [CrossRef]
19. Meinzer, O.E. *The Occurrence of Ground Water in the United States with a Discussion of Principles*; US 89 Survey Water-Supply Paper No. 489; U.S. Department of the Interior: Washington, DC, USA, 1923. [CrossRef]
20. Springer, A.E.; Stevens, L.E. Spheres of discharge of springs. *Hydrogeol. J.* **2009**, *17*, 83–93. [CrossRef]
21. Kresic, N. Types and classifications of springs. In *Groundwater Hydrology of Springs: Engineering, Theory, Management, and Sustainability*; Kresic, N., Stevanovic, Z., Eds.; Elsevier: Amsterdam, The Netherlands, 2010; pp. 31–86. ISBN 978-1-85617-502-9.
22. Botosaneanu, L. (Ed.) *Studies in Crenobiology: The Biology of Springs and Springbrooks*; Backhuys Publishers: Leiden, The Netherlands, 1998; ISBN 90-73348-04-8. [CrossRef]
23. Dumnicka, E.; Galas, J. An overview of stygobiontic invertebrates of Poland based on published data. *Subterr. Biol.* **2017**, *23*, 1–8. [CrossRef]
24. Pipan, T.; Culver, D.C. The unity and diversity of the subterranean realm with respect to invertebrate body size. *J. Cave Karst Stud.* **2017**, *79*, 1–9. [CrossRef]
25. Halse, S. Research in calcretes and other deep subterranean habitats outside of caves. In *Cave Ecology*; Moldovan, O.T., Kováč, L., Halse, S., Eds.; Springer Nature: Gland, Switzerland, 2018; pp. 415–434. ISBN 978-3-319-98850-4.

26. Humphreys, W.F. Groundwater calcrete aquifers in the Australian arid zone: The context of an unfolding plethora of stygal biodiversity. *Rec. West. Aust. Mus.* **2001**, *64*, 63–83. [CrossRef]
27. Pentecost, A. *Travertine*; Springer: Berlin/Heidelberg, Germany, 2005; ISBN 978-1-4020-3523-4.
28. Leruth, R. La biologie du domaine souterrain et la faune cavernicole de la Belgique. *Mém. Mus. Hist. Nat. Belgique* **1939**, *87*, 1–506.
29. Howarth, F.G.; Moldovan, O.T. Where cave animals live. In *Cave Ecology*; Moldovan, O.T., Kováč, L., Halse, S., Eds.; Springer Nature: Gland, Switzerland, 2018; pp. 41–68. ISBN 978-3-319-98850-4.
30. Poulson, T.L. The Mammoth Cave ecosystem. In *The Natural History of Biospeleology*; Camacho, A.I., Ed.; Museo Nactional de Ciencias Naturales: Madrid, Spain, 1992; pp. 569–612. ISBN 84-00-07280-4.
31. Meštrov, M. Un nouveau milieu aquatique souterrain: Le biotope hypotelminorheique. *C. R. Acad. Sci.* **1962**, *254*, 2677–2679.
32. Keany, J.M.; Christman, M.C.; Milton, M.; Knee, K.L.; Gilbert, H.; Culver, D.C. Distribution and structure of shallow subterranean aquatic arthropod communities in the parklands of Washington, D.C. *Ecohydrolgy* **2018**, *12*, e2044. [CrossRef]
33. Culver, D.C.; Trontelj, P.; Zagmajster, M.; Pipan, T. Paving the way for standardized and comparable subterranean biodiversity studies. *Subterr. Biol.* **2012**, *10*, 43–50. [CrossRef]
34. Orghidan, T. Un nou domeniu de viata acvatica subterana: "Biotopul hiporeic". *Buletin Stiintific Sectia de Biologie si Stiinte Agricole si Sectia de Geologie si Geografie* **1955**, *7*, 657–676.
35. Käser, D.H. A new habitat of subsurface waters: The hyporheic biotope, by Traian Orghidan (1959). *Fundam. Appl. Limnol.* **2010**, *176*, 291–302.
36. Malard, F.; Tockner, K.; Dole-Oliver, M.-J.; Ward, J.V. A landscape perspective of surface-subsurface hydrological exchanges in river corridors. *Freshw. Biol.* **2002**, *47*, 621–640. [CrossRef]
37. Dumnicka, E.; Galas, J.; Krodkiewska, M.; Pociecha, A. The diversity of annelids in subterranean waters: A case study from Poland. *Knowl. Manag. Aquat. Ecosyst.* **2020**, *421*, 16. [CrossRef]
38. Hahn, H.J.; Fuchs, A. Distribution patterns of groundwater communities across aquifer types in south-western Germany. *Freshw. Biol.* **2009**, *54*, 848–860. [CrossRef]
39. Culver, D.C.; Pipan, T. Superficial subterranean habitats-gateway to the subterranean realm? *Cave Karst Sci.* **2008**, *35*, 5–12.
40. Blatnik, M.; Culver, D.C.; Gabrovšek, F.; Knez, M.; Kogovšek, B.; Kogovšek, J.; Liu, H.; Mayaud, C.; Mihevc, A.; Mulec, J.; et al. Changing perspectives on subterranean habitats. In *Karstology in the Classical Karst*; Knez, M., Otoničar, B., Petrič, M., Pipan, T., Slabe, T., Eds.; Springer: Cham, Switzerland, 2020; pp. 183–205. ISBN 978-3-030-26827-5.
41. Bradford, T.; Adams, M.; Humphreys, W.F.; Austin, A.D.; Cooper, S.J.B. DNA barcoding of stygofauna uncovers cryptic amphipod diversity in a calcrete aquifer in Western Australia's arid zone. *Mol. Ecol. Resour.* **2010**, *10*, 41–50. [CrossRef]
42. Brancelj, A.; Boonyanusith, C.; Watiroyram, S.; Senamuang, L.O. The groundwater-dwelling fauna of Southeast Asia. *J. Limnol.* **2013**, *72*, 327–344. [CrossRef]
43. Nola, M.; Togouet, S.H.Z.; Marmonier, P.; Kayo, R.T.; Piscart, C. An annotated checklist of freshwater stygobiotic crustaceans of Africa and Madagascar. *Crustaceana* **2012**, *85*, 1613–1631. [CrossRef]
44. Ríos-Escalante, P.; Parra-Coloma, L.; Peralta, M.A.; Pérez-Schultheiss, J.; Rudolph, E.H. A checklist of subterranean water crustaceans from Chile (South America). *Proc. Biol. Soc. Wash.* **2016**, *129*, 114–128. [CrossRef]
45. Culver, D.C.; Holsinger, J.R.; Feller, D.J. The fauna of seepage springs and other shallow subterranean habitats in the mid-Atlantic Piedmont and Coastal Plain, U.S.A. *Northeast. Nat.* **2012**, *19*, 1–42. [CrossRef]
46. Trajano, E.; Bichuette, M.E. Diversity of Brazilian subterranean invertebrates, with a list of troglomorphic taxa. *Subter. Biol.* **2010**, *7*, 1–16.
47. Holsinger, J.R.; Baroody, R.A.; Culver, D.C. *The Invertebrate Cave Fauna of West Virginia*; Bulletin 7 WV Speleological Survey; West Virginia Speleological Survey: Barrackville, WV, USA, 1976.
48. Gibert, J. Protocols for the assessment and conservation of aquatic life in the subsurface (PASCALIS), a European project. In *Mapping Subterranean Biodiversity/Cartographie de la Biodiversité Souterrain*; Culver, D.C., Deharveng, L., Gibert, J., Sasowsky, I., Eds.; Karst Waters Institute Special Publication 6; Karst Waters Institute: Charles Town, WV, USA, 2001; pp. 19–21. ISBN 0-9640258-5-x.

49. Deharveng, L.; Stoch, F.; Gibert, J.; Bedos, A.; Galassi, D.; Zagmajster, M.; Brancelj, A.; Camacho, A.; Fiers, F.; Martin, P.; et al. Groundwater biodiversity in Europe. *Freshw. Biol.* **2009**, *54*, 709–726. [CrossRef]

50. Eme, D.; Zagmajster, M.; Delić, T.; Fišer, C.; Flot, J.-F.; Konecny-Dupré, L.; Pálsson, S.; Stoch, F.; Zakšek, V.; Douady, C.J.; et al. Do cryptic species matter in macroecology? Sequencing European groundwater crustaceans yields smaller ranges but does not challenge biodiversity determinants. *Ecography* **2017**, *40*, 424–436. [CrossRef]

51. Zagmajster, M.; Eme, D.; Fišer, C.; Galassi, D.; Marmonier, P.; Stoch, F.; Cornu, J.-F.; Malard, F. Geographic variation in range size and beta diversity of groundwater crustaceans: Inputs from habitats with low thermal seasonality. *Glob. Ecol. Biogeogr.* **2014**, *23*, 1135–1145. [CrossRef]

52. Bou, C.; Rouch, R. Un nouveau champ de recherches sur la faune aquatiques souterraine. *C. R. Acad. Sci. Paris* **1967**, *265*, 369–370.

53. Malard, F.; Dole-Olivier, M.-J.; Mathieu, J.; Stoch, F. *Sampling Manual for the Assessment of Regional Groundwater Biodiversity*; European Project PASCALIS: Lyon, France, 2002.

54. Pipan, T. *Epikarst-A Promising Habitat*; Založba, ZRC: Ljubljana, Slovenia, 2005; ISBN 961-6500-90-2.

55. Pipan, T.; Holt, N.; Culver, D.C. How to protect a diverse, poorly known, inaccessible fauna: Identification and protection of source and sink habitats in the epikarst. *Aquat. Conserv. Mar. Freshw. Ecosyst.* **2010**, *20*, 748–755. [CrossRef]

56. Niemiller, M.L.; Porter, M.L.; Keany, J.; Gilbert, H.; Culver, D.C.; Fong, D.W.; Hobson, C.S.; Kendall, K.D.; Taylor, S.J. Evaluation of eDNA for groundwater invertebrate detection and monitoring: A case study with endangered *Stygobromus* (Amphipoda: Crangonyctidae). *Conserv. Genet. Resour.* **2017**, *10*, 247–257. [CrossRef]

57. Dumnicka, E.; Galas, J.; Najberek, K.; Urban, J. The influence of Pleistocene glaciations on the distribution of obligate aquatic subterranean invertebrate fauna in Poland. *Zool. Anz.* **2020**, *286*, 90–99. [CrossRef]

58. Malard, F.; Boutin, C.; Camacho, A.I.; Ferreira, D.; Michel, G.; Sket, B.; Stoch, F. Diversity patterns of stygobiotic crustaceans across multiple spatial scales in Europe. *Freshw. Biol.* **2009**, *54*, 756–776. [CrossRef]

59. Culver, D.C.; Deharveng, L.; Bedos, A.; Lewis, J.J.; Madden, M.; Reddell, J.R.; Sket, B.; Trontelj, P.; White, W. The mid-latitude biodiversity ridge in terrestrial cave fauna. *Ecography* **2006**, *29*, 120–128. [CrossRef]

60. Culver, D.C.; Sket, B. Hotspots of subterranean biodiversity in caves and wells. *J. Cave Karst Stud.* **2000**, *62*, 11–17.

61. Culver, D.C.; Pipan, T. *The Biology of Caves and Other Subterranean Habitats*; Oxford University Press: Oxford, UK, 2019; ISBN 978-0-19-882076-5.

62. Zagmajster, M.; Malard, F.; Eme, D.; Culver, D.C. Subterranean biodiversity patterns from global to regional scales. In *Cave Ecology*; Moldovan, O.T., Kováč, L., Halse, S., Eds.; Springer Nature: Gland, Switzerland, 2018; pp. 195–228. ISBN 978-3-319-98850-4.

63. Guzik, M.T.; Austin, A.D.; Cooper, S.J.B.; Harvey, M.S.; Humphreys, W.F.; Bradford, T.; Eberhard, S.M.; King, R.A.; Leys, R.; Muirhead, K.A.; et al. Is the Australian subterranean fauna uniquely diverse? *Invertebr. Syst.* **2010**, *24*, 407–418. [CrossRef]

64. Knight, L.R.F.D. The aquatic macro-invertebrate fauna of Swildon's Hole, Mendip Hills, Somerset, UK. *Cave Karst Sci.* **2011**, *38*, 81–92.

65. Pipan, T.; Culver, D.C.; Papi, F.; Kozel, P. Partitioning diversity in subterranean invertebrates: The epikarst of Slovenia. *PLoS ONE* **2018**, *13*, e0195991. [CrossRef] [PubMed]

MDPI
St. Alban-Anlage 66
4052 Basel
Switzerland
Tel. +41 61 683 77 34
Fax +41 61 302 89 18
www.mdpi.com

*Water* Editorial Office
E-mail: water@mdpi.com
www.mdpi.com/journal/water

9 783036 526232